S0-AEA-300

AMERICA ON THE BRINK

The Next Added 100 Million Americans

FROSTY WOOLDRIDGE

authorHOUSE®

AuthorHouse™
1663 Liberty Drive
Bloomington, IN 47403
www.authorhouse.com
Phone: 1-800-839-8640

© 2009 Frosty Wooldridge. All rights reserved.

No part of this book may be reproduced, stored in a retrieval system, or transmitted by any means without the written permission of the author.

First published by AuthorHouse 4/20/2009

ISBN: 978-1-4389-6074-6 (sc)

Printed in the United States of America
Bloomington, Indiana

This book is printed on acid-free paper.

This book honors all the unsung citizens in countries around the world that see our dilemma and work toward restoring a sustainable planet for all humans, plants and animals. Your courageous actions drive possibilities. I feel profound gratitude for your tireless efforts, creative energy and sheer determination in the face of astounding ignorance, apathy and opposition.

TABLE OF CONTENTS

SECTION VI: GROWTH, DOUBLE-EDGED SWORD

SECTION VII: ACTIONS AND SOLUTIONS

FOREWORD

Across the planet, humanity faces its greatest crisis in the 21st century created by its own hyper-population growth. Worldwide demographic reports show human beings growing from 6.7 billion to a mind-numbing 9.2 billion low estimate to a high of 9.8 billion in 40 years. Whether by the hands of culture or religion, human beings propagate beyond any other species on Earth.

Via population overload, humanity faces its greatest dilemma since the beginning of recorded history. Over the ages, we survived wars, diseases and pestilence, but our survival never depended on our ability to stop proliferating beyond Earth's carrying capacity.

Few citizens in first world countries possess an inkling of the raging struggle for survival going on around the planet for a third of the human race.

Meanwhile, Americans remain unaware of an appalling statistic facing their children: half of all growth on the planet by 2050 explodes from eight nations--India, China, the United States, Pakistan, Ethiopia, Bangladesh, Nigeria and the Democratic Republic of the Congo.

As a result of accelerating population growth around the globe, first world countries find themselves besieged by immigrants flooding from third world countries at gathering speeds. No matter how many migrate, the world grows by a net gain of 77 million human beings annually.

In the Western Hemisphere, Mexico expects to grow from 106 million to 153 million in 40 years. If the United States continues its endless immigration policies, its citizens face ominous consequences from water shortages, lack of energy and food scarcity. Additionally, environmental calamities across the spectrum explode in a virtual Pandora's Box of predicaments.

These realities—both by the demographic company we keep and the implications for the United States—might turn the reader an Earth Day green.

America stands at the most critical juncture of its 233 year history since the signing of the Constitution. What dilemma faces our nation?

Hyper-population growth: Plague of 21st century planet Earth

The world grows more constricted by adding 9,000 newborns per hour net gain over deaths, 216,000 per day and well over 77 million added human beings per year habitually in third world countries. According to a recent Time Magazine report, eight million adults starve to death annually around the globe.

Additionally, every year, 10 million children under age 12 die of starvation or starvation related diseases. All mankind faces this "human dilemma."

However, world leaders continue ignoring it. Therefore, refugees escape misery, starvation and hopelessness found in their own countries. Where do they flee as their main destination?

Not China! Not India! Not Bangladesh! Not Africa!

Millions rush to functioning republics and democracies in Western Europe. France now endures five million immigrants. Multitudes rush over borders in Germany, Holland, Norway, Denmark, Spain, Austria, Sweden, England and Italy. In the Western Hemisphere, the third world's poor, uneducated and unskilled stream into America at 1.0 million illegal migrants per year while 1.2 million arrive via legal avenues. Canada accepts millions of refugees. The quandary however: the line never ends. Again, this procession swells by 77 million annually.

In our short-sightedness, we fail to address the obvious alternative to increase our assistance to indigent citizens in their own nations. Instead, we provide massive welfare and job magnets in America.

In his prophetic book, *Camp of the Saints*, French author Jean Raspail featured a ragged fleet of ships sailing away from India where the people had destroyed their civilization by their own overpopulation. However, instead of solving their own problems, they set sail toward France.

"You don't know my people—the squalor, superstitions, the fatalistic sloth that they've wallowed in for generations. You don't know what you're in for if that fleet of brutes ever lands in your lap. Everything will change in this country of yours. They will swallow you up."

One look at California's hospitals, schools, crime, diseases and social chaos sobers any American as to the speed of this population juggernaut. We await accelerating consequences in more aspects than simple numbers. California's pandemonium in its hospitals, newly virulent diseases and massive debt clearly point to the doorstep of perpetual population growth.

Frantic migrants from around the world invade America's borders at breakneck speed. Does that solve problems in their countries? No! It temporarily solves individuals' problems, but they too will see the luster quickly diminish as America exceeds its carrying capacity.

Ironically, no one addresses the source of this state of affairs. Not one U.S. Senator, state governor, President Obama or any high ranking U.S. leaders address the population equation. It remains the 'sacred cow' of American politics. Additionally, the media remains complicit in its silence. "In our race for survival," said Richard Heinberg, "our media people are the cheerleaders for our extinction."

Consequently, immigration forces the United States to the position of third fastest growing country in the world. We expect to double our current 306 million population in 70 years. This effectively doubles our air pollution, auto and air traffic, gridlock, size of cities, carbon footprint, water usage, species extinction, use of non-renewable resources, sprawl and all other negative factors related to hyper-population growth.

What are you doing about it? Complaining? Not sure how to take action? Are you pressing your remote to the latest rerun of "Friends" to numb your mind or feel good? Maybe it will go away on its own if you ignore it long enough. Wrong!

"The problems in the world today are so enormous they cannot be solved with the level of thinking that created them." Einstein

Riding a demographic Titanic

Passengers on the Titanic did not know their fate. We do! We know we cannot allow this civilization to continue in the same direction—heading toward the population iceberg. We must make our voices heard all the way up to the captain and crew. This country is our ship. America holds the future for our families and children. We demand this president and Congress change course.

As we plunge into the early years of the 21st century, I'll admit to you: our civilization stands nostril-deep in trouble. Even if we closed our borders today to all immigration, our 'population momentum' would add 35 million Americans in 40 years before stabilizing within our country. Sustaining that added population load will prove gargantuan enough! The projected 100 million via immigration in 30 years will lift solutions beyond our reach. Finally, Mother Nature cares little about anyone's race, creed or color when it comes to excessive human numbers and limited carrying capacity. Not only humans, but all living creatures become ensnarled in this dilemma.

My own bicycle expeditions over 100,000 miles across six continents, from the Arctic to Antarctica and six times across the USA coast to coast, afforded me profound visual experiences of what's coming to the United States and Canada. While pedaling through Asia from the Wall of China to India, I witnessed horrific human degradation via hyper-population growth. Cycling the entire perimeter of Australia and the length of South America gave me firsthand knowledge of humanity's impact on this planet. My cycling journey from Nord Cap, Norway, 600 miles north of the Arctic Circle, to Athens, Greece provided sobering knowledge of human excesses.

As a math and science teacher in the 1970s, I felt a pressing responsibility to educate and prepare my students for their participation in this civilization. Today, still feeling a deep responsibility to all humanity, I wrote this book as an educational instrument and prod to create discussion and debate on our most pressing crisis in the 21st century: overpopulation in America, Canada and globally.

Yes, we face a "human dilemma" on a worldwide stage. It grows worse by the day. It appears daunting. Nonetheless, like the founders of this republic, we must move forward with purpose, passion and intentions. We start with ourselves: form groups, work in our communities, expand to the state level, connect with those in other states, push toward national exposure and work our way into the highest political arenas. Our main intention: change this nation's course from its current self-destructive path—to one of a viable, sustainable and peaceful future.

This book educates and guides you on how to become part of the solution instead of ineffectually complaining about the problem. First, it will connect the dots by showing you what happens in every sector of the United States as well as the world. Second, it steers you toward effective participation.

Who is your hero? Your teacher? Your parents? John Muir? Thomas Jefferson? Betsy Ross? LeBron James? Dr. Martin Luther King? Barack Obama? Susan B. Anthony? Cesar Chavez? Teddy Roosevelt? Barbara Jordan? Michelle Obama? Michael Jordan? Eleanor Roosevelt? Gandhi? Mandela? Charles Lindbergh? Federico Pena? Lupe Moreno?

Those regular men and women display(ed) uncommon determination. They thrived in their time; this is your moment. Take their energy, creative talents and drive—and make those qualities your own. You possess the same abilities, passion and personal power to change history toward a brighter future.

Conviction without action proves worthless. Passion coupled with action changes history. Let's get busy.

Frosty Wooldridge
www.numbersusa.com
www.thesocialcontract.com
www.fairus.org
www.frostywooldridge.com
Louisville, CO

This grand show is eternal. It is always sunrise somewhere; the dew is never all dried at once; a shower is forever falling; vapor is ever rising. Eternal sunrise, eternal sunset, eternal dawn and gloaming, on sea and continents and islands, each in its turn, as the round earth rolls.

John Muir

SECTION I: OVERVIEW

CHAPTER 1: CONSEQUENCES
OF A HUMAN KATRINA

---

*"The raging monster upon the land is population growth. In
its presence, sustainability is but a fragile theoretical construct.
To say, as many do, that the difficulties of nations are not due
to people but to poor ideology and land-use management is
sophistic."*

Harvard scholar and biologist E.O. Wilson

According to Katie Couric, Brian Williams and Charles Gibson,
the United States surpassed 300,000,000 people in October 2006.
The U.S. Census Bureau, based on accelerating growth levels, shows
America adding 100 million[1] people by 2035.

For those asleep at the wheel—that's three decades from now—a
blink in time.

To place this kind of horrific growth rate into perspective, it
resembles a "Human Tsunami." Much like nature's earthquakes that
occurred beneath the ocean on December 26, 2004 in Sri Lanka,
the energy wave sped under the surface for hours and hundreds of
miles without notice. Once it hit the shoreline, it created cataclysmic
devastation and tens of thousands of deaths. Why? No one suspected
it. None took action to save themselves. They didn't know it raced
toward them. Once it hit, everyone became victims! That tsunami
rendered human tragedy of epic proportions!

Fast forward to 2005, another kind of natural disaster erupted
in the Gulf of Mexico. Hurricane Katrina approached the U.S. Gulf
Coast featuring 200 mile per hour winds. For a week before it hit,
everyone witnessed swirling clouds funneling high in the heavens
on their TV sets. Five different kinds of people reacted to the news
reports:

1. People assessed the danger and drove any kind of vehicle they could procure out of the hurricane area. They acted on their knowledge and proved proactive.
2. Some said, "We've ridden out these hurricanes before so we can ride this one out, too." They thought from the past and acted as if history would serve them in the present. Some survived and some died. Thinking in the past!
3. Others said, "Let's wait and see!" Those people suffered traffic jams, death, burning busses, gas stations without fuel and miles of Interstates jammed with the same kind of people. Some died and some survived. Pot luck thinking!
4. Still others said, "Maybe it will change course or drop down to a Category "1" storm." *'Maybe'* came with a high price tag. They became victims or survivors. They resorted to chance rather than personal actions!
5. The poor could not drive away or afford transport. They became victims beyond their own choices.

A Human Katrina slams America

In 1963, the United States featured a stable population of 194 million people. America accepted 170,000 persons as the annual legal immigration rate. At the same time, that many people departed the United States annually. Births and deaths matched one another. Thus—a stable population! However, change burst upon the scene.

In 1965, the U.S. Senate passed the "Immigration Reform Act" shepherded by Senator Teddy Kennedy. Lyndon Baines Johnson signed it without fanfare—jumping immigration rates to 1.1 and as high as 1.4 million annually.

Within 41 years, the United States added 106 million people through October 2006. That new benchmark set the stage for adding 100 million more by 2035.[2] How? Moving into the 21st century, not only legal, but massive illegal immigration pours over U.S. borders.

Accelerating immigration

Time Magazine's feature story on September 20, 2004, "*What Happened*?" estimated three million people crossing into America illegally every year.

The U.S. Senate in June 2007 attempted passage of S.B. 1369 that would have doubled current immigration levels from 1.2 million to 2.4 million annually. It increased work visas by tens of thousands. It continued allowances for millions in *chain migration*. It allowed millions more in *anchor babies* which constitute children of persons that entered the United States illegally. The senate bill did nothing to stop illegal immigration. The bill failed five times, while senators continued pressing for greater population influx into the United States.

American citizens face a challenging future

What about overloaded cities? Overwhelmed schools? How about water, farm land, energy, air quality, food sources, species habitat, and dozens of other issues?

What do we as a nation face if we allow this "Human Katrina" to crash upon our shores? It negatively affects every aspect of our society: environment, sustainability, culture, language and viability as a civilization.

Unfortunately, no national leader promotes a strategic population initiative. That leaves our civilization at risk and at the whim of a future explored by Pulitzer Prize winner Jared Diamond in his book: *COLLAPSE: How Societies Choose to Fail or Succeed*. As a leading anthropologist, he discovered why civilizations of the past fell into ruin.

Dilemma of the 21st century

Dr. Albert Bartlett of the University of Colorado asked the most prominent question of the 21st century: "Can you think of any problem in any area of human endeavor on any scale, from microscopic to global, whose long-term solution is in any demonstrable way aided,

3

assisted, or advanced by further increases in population, locally, nationally or globally?"

In 1900, the world population reached 1.6 billion; today, it exceeds 6.7 billion. By mid century experts expect world population to grow to a low of 9.2 to as high as 9.8 billion. (Source: Population Reference Bureau)

According to March 14, 2005 Time Magazine, eight million people already starve to death around the planet annually. Over 35 percent of humanity lacks clean drinking water. Species extinction exceeds thousands annually.

What do we hope to accomplish by adding another 3.0 billion people to the planet with consequences already crushing us with current population levels?

In the next 50 years, if we continue on this path, you can expect 1.0 million to as many as 3.0 million more people added to your state depending on location. Once their numbers manifest, they won't vanish. Why? Because, at the same time, that 100 million people spread to other states! Texas adds 12 million by 2025, while Arizona adds four million and California adds 20 million by 2035.

Finite carrying capacity

Today, states like Colorado, New Mexico, Arizona and California don't possess enough water for their residents. Aquifers degrade as fast as diesel engines pump them dry. No matter how many reservoirs we build, nature won't rain or snow more to accommodate added millions of people. Water shortages and rationing will become the norm while green lawns become as extinct as five cent candy bars.

At the same time, states like Colorado lost 1.65 million acres of prime farm land to development in the past 15 years. Why? Adding 1.3 million people requires homes, roads, malls and commercial buildings. Colorado will add five million more by 2050. Latest reports show Colorado losing 3.1 million more acres by 2022. How many by 2050? An additional 2.0 million more acres will be placed under concrete and asphalt via roads, malls and housing. Ask yourself: have you ever seen cows grazing on concrete or corn growing out of pavement? What will be the acreage losses in your state? Your losses will be commensurate to your population growth.

While we exceed the land's carrying capacity, we squeeze ourselves in like sardines with "smart growth" and "slow growth" and "managed growth." Any way you stack it, growth adds vehicles, homes, power plants, malls and smoke stacks. Have you noted increased traffic in your city? Have you seen the Brown Cloud thicken in toxicity over your area? How about the bumper-to-bumper traffic?

How about your quality of life? Standard of living? What about species extinction? Air pollution? Acid rain? Crowding of national parks? Lakes? Streams? How about soil erosion?

We drive this population train toward the edge of the Grand Canyon—blind, deaf and clueless. Lacking a strategic plan, commensurate vision statements and a full-set of first-class actions—we swell ourselves into a world of hurt. We can take action by applying the brakes now, but once we career over the edge, we become powerless.

When will local and national leaders speak out? When will newspapers, TV and radio talk shows deal with our number one crisis: population? For a hint of our future, visit www.thesocialcontract.com for the Social Contract Quarterly.

Of the five types of choices people made concerning Hurricane Katrina, as a nation, what choices will we make? Take action now? Wait and see? Think we will ride it out again like we rode it out before? Pray for this population juggernaut to change course? Or, resign and become victims for certain?

Each of us rides on Spaceship Earth. Perhaps Chief Seattle said it best, "All things and all peoples are connected."

In the end, our children and future generations walk into the cross hairs of our poor choices today.

This book invites you to educate yourself with critical thinking, standing tall—and then, taking action. Your children and future residents of America will look back and thank you for bequeathing them a sustainable civilization and planet.

Their future lies in your hands today.

CHAPTER 2: OUR HUMAN DILEMMA

"Pressures resulting from unrestrained population growth put demands on the natural world that can overwhelm any efforts to achieve a sustainable future. If we are to halt the destruction of our environment, we must accept limits to that growth."

World scientists' warning to humanity, signed by 1600 senior scientists from 70 countries including 102 Nobel Prize laureates, November 18, 1992

By 2050, at the current rate of growth, our country adds another 138 million to reach a total of 438 million. Many experts expect higher numbers. Is it a milestone or millstone?

In the face of scientific evidence of melting polar ice caps, accelerating species extinction, water shortages, soil erosion, air pollution, acid rain and vanishing farmland—where do we find a national leader to address America's most ominous dilemma early in the 21st century? Few leaders emerge in the political, educational or religious realm.

While the Pope witnesses starvation, misery and suffering worldwide, he promotes maximum human birth rates. Islam commits to the same agenda as do most other faiths. Religious leaders refuse to budge from their 2000 year old dogmas. Their clout still trumps the masses' ability or willingness to think critically. Those leaders avoid the realities of the 21st century.

U.S. industrial giants won't speak about it. Most Americans ignore its reality. A few notables like Dr. John Tanton, Roy Beck, Sharon Stein, Paul Ehrlich, Dr. Diana Hull, David Paxson, Bill Ryerson, Dan Stein, Marilyn Hempel, Buck Young, Kathleene Parker, Garrett Hardin, Dr. Albert Bartlett of Colorado University and former Colorado Governor Richard D. Lamm—educate Americans about their future population calamity.

But, like the Amtrak Express on the midnight run, it's comin' and it's comin' fast

Silent-assertion! One hundred-fifty years ago, one of my favorite authors, Mark Twain said, "Almost all lies are acts, and speech has no part in them. I am speaking of the lie of silent assertion; we can tell it without saying a word. For example: it would not be possible for a humane and intelligent person to invent a rational excuse for slavery; yet you will remember that in the early days of emancipation agitation in the North, the agitators got but small help or countenance from anyone. Argue, plead and pray as they might, they could not break the universal stillness that reigned, from the pulpit and press all the way down to the bottom of society—the clammy stillness created and maintained by the lie of silent assertion; the silent assertion that there wasn't anything going on in which humane and intelligent people were interested.

"The universal conspiracy of the silent assertion lie is hard at work always and everywhere, and always in the interest of a stupidity or sham, never in the interest of a thing fine or respectable. It is the most timid and shabbiest of all lies…the silent assertion that nothing is going on which fair and intelligent men and women are aware of and are engaged by their duty to try to stop."

In Twain's time, slavery continued as the silent-assertion of the day until it imploded into states' rights and the Civil War.

Today, in Congress and the White House, you witness a complete abrogation of common sense, action for the common good and visionary engagement toward the future. With each new scandal, aberrant side-dishes surface weekly that detract from the harsh realities we face.

"Have you heard the latest on Britney Spears?" gossip show hosts lament. "Oh, my gosh! What about Angelina Jolie's affair with Jodie Foster, or was that Paris Hilton or was it with a turtle?"

Reality check!

America faces unending migrant millions from its southern border. In the last century, Mexico expanded from 50 million poverty stricken peasants to 106 million today. By mid century, because of, or due to, religious and cultural propensities, Mexico expects 153 million people. If you think they engage solutions to their problems, think again.

Not only Mexico, but millions in South America, Africa, India, China, Bangladesh and the Middle East explode out of their demographic as well as ecological britches and carrying capacity limits.

We can ignore reality for a limited amount of time, but reality will not ignore us

As this book attests, no first world country can escape the inevitable migration of millions from overloaded, overcrowded and environmentally unsustainable countries around the world. To avoid a perilous future, first world as well as third world countries must take action immediately, profoundly and with tenacious determination!

You may recall a movie starring Will Smith titled, "Independence Day" whereby an alien force invaded planet Earth. All nations came together to fend off the invader.

Can we unite humanity in this 21st century to act in unison for our own survival? Do we realize the 'enemy' proves to be our own fecundity? Can we unite to change our cultural and biological propensities away from endless growth?

We better! As Henry Kendall said, "We can bring about population stabilization graciously or nature will do it brutally."

What we face

If I could take you for a two week trip to Mexico City, Mexico; Shanghai, China; Bombay, India; Dhaka, Bangladesh; or Dakar, Egypt—you would grow sick to your stomach. You would be inspired to take action. You wouldn't want your children to live with the misery those people endure, by the millions, as they cling to life—every day of their lives.

The PBS journalist Bill Moyers asked the great science fiction writer Isaac Asimov, "What happens to the idea of the dignity of the human species if this population growth continues at its present rate?"

Asimov replied, "It will be completely destroyed. I use what I call the bathroom metaphor: if two people live in an apartment and there are two bathrooms, then both have freedom of the bathroom. You can go to the bathroom anytime you want to stay as long as you like for whatever you need.

"But if you have twenty people in the apartment and two bathrooms, no matter how much every person believes in freedom of the bathroom, there is no such thing. You have to set up times for each person; you have to bang on the door, 'Aren't you done yet?'"

He concluded, "In the same way, democracy cannot survive overpopulation. Human dignity cannot survive. Convenience and decency can't survive. As you put more and more people onto the world, the value of life not only declines, it disappears. It doesn't matter if someone dies, the more people there are, the less one person matters."

How to beat nature to the punch

First of all, we must educate ourselves. Second, we must take action.

Each chapter of this book exposes ramifications we face as to social, cultural, environmental and planetary results of continued growth in America and around the world. Once you become highly astute on what we face, you will be directed to personal, local, national and international web sites in order to take action.

Above all, bring your creative ideas, mastermind groups and networking toward a sustainable future for America and planet Earth.

CHAPTER 3: BOW OF THE TITANIC

> "Captain Edward John Smith steamed the Titanic into the iceberg-filled North Atlantic as if he were cruising through the Bahamas. History tells us he made a grave mistake."
>
> FHW, environmentalist

At one point, California boasted itself the most beautiful state in the Union. In 1950, it housed a reasonable 10 million people. Known as the land of milk and honey—California's mountains, coastline and weather beckoned. California condors soared through limitless blue skies. Yosemite National Park, giant Sequoia redwoods, whales and seals along its coastline, Hollywood and 77 Sunset Strip—created the California mystique!

Fifty-seven years later, 37.5 million people cram, jam, gridlock and fume in their fumes on 'forever' crowded freeways. Growing at 1,700 people daily and over 600,000 annually—California expects an added 21 million people within 35 years. (Source: www.capsweb.org)[3]

To illustrate 'environmental refugees', we discover 40 percent of Los Angeles residents were born outside the U.S. (Source: www.cis.org) They arrived from Mexico, Korea, China, Central, South America and Asia.

Result? Massive subdivision housing sprawl! Roads, malls, schools, churches, firehouses and homes devour land like Kansas wheat combines. Developers demolish nature. They guzzle water. They vomit black smoke into the air. Cars whiz around like mad hornets. The more compacted the traffic, the more drivers suffer 'road rage'. Few smiling faces can be seen on California freeways! Drivers busy themselves trying to stay alive.

Joe Guzzardi, a writer and college professor in Lodi, California, recently moved to Pennsylvania, said, "If we continue our suicidal immigration path, whether the inevitable development takes the form of sprawl by building on a city's periphery or landfill by building inside the city limits, the net result will be the same—an eroded quality of life and a vanished sense of place."

California's developers brag 'smart growth', however, whether that means 'slow growth', 'managed growth', 'brilliant growth', 'dumb growth', 'fast growth', or 'snail's pace growth'—it equals up to 30 million more people swarming all over California by mid century. (Source: Fogel/Martin, March 2006, "US Population Projections")[4]

Governor Schwarzenegger and state treasurer Phil Angelides stuff themselves into the pockets of developers. Angelides said, "We are a state of 26 million cars, SUVs and trucks that travel 314 billion miles a year and burn 15 billion gallons of gas. We are on a path over the next 20 years to become a state with 36 million cars that travel 446 billion miles and burn nearly 18 billion gallons. We must choose to grow smarter, to give Californians more transportation options, the choice to drive fewer miles and burn fewer gallons of fossil fuel."

That sounds like an idiot talking to a moron who then relays the story to an imbecile! Even Goober on "Mayberry RFD" possessed more common sense!

Some choice! How intelligent is Angelides' statement? To top it off, former President Bush, in his State-of-the-Union speech said, "In the next 10 years by 2017, the United States will reduce oil consumption by 20 percent by using conservation, hybrid cars and ethanol."

He forgot to report America adding 30 million people in that 10 year span. Therefore, our consumption can only rise by a factor of 30 million people using gas, coal, natural gas and wood for energy.

Journalist Joe Guzzardi said, "If people would contemplate the additional 100 million people coming our way in the not too distant future, and our current gluttonous land use, then they might become more alarmed. In a word, the problem is population. If it can be stabilized through sensible immigration policies, then we have a chance to level off growth. We'd have a chance to save our state and the United States."

This journalist has bicycled the length and width of California four times in the past 25 years. I've seen it change from paradise to hell on earth. Too many people fill its parks with too much trash. Its ocean beaches suffer dying seals and seabirds from too much sewage, plastic, glass and aluminum pollution. As Katie Couric on CBS reported, fish stocks dropped 90 percent in the past decade. California skies fill with toxic smoke too thick to breathe. Yosemite National Park suffers wall-to-wall crowding. Millions of cars create a kind of insanity of movement far removed from the natural world. Condors no longer soar in pristine skies because the last of them perch in cages built to save their species.

Constant tension fills places like Los Angeles and San Francisco. You cannot avoid the crowding, metal, concrete, glass, wires, buildings, roads and loss of sense of place.

One of my favorite writers, a Californian, John Muir said, "Tell me what you will of the benefactions of city civilization, of the sweet security of streets—all as part of the natural up-growth of man towards the high destiny we hear so much of. I know that our bodies were made to thrive only in pure air, and the scenes in which pure air is found. If the death exhalations that brood the broad towns in which we so fondly compact ourselves were made visible, we should flee as from a plague. All are more or less sick; there is not a perfectly sane man in all of San Francisco." (September 1874, JOM, page 191-92.)

If the United States can be compared to the Titanic, our country steams into dangerous waters, much too fast and overloaded with too many people to stay afloat. California might be the bow of our ship and, as it begins failing, its own 'environmental refugees' cannot help but abandon ship like rats in a hurricane. Had the Titanic been able to stop the in-flooding of the North Atlantic, it would not have become the greatest seagoing catastrophe of the last century.

However, few want to speak up or take action. I am confounded that no national leaders step into the center ring to call for a national population policy. None talk about stopping the in-flooding of humanity with the simple choice of reasoned action.

It didn't make any difference on the Titanic if you were first class, third class or shoveling the coal in the boilers. When the ship sank, everyone became a victim in one form or another. As California fails in areas of water shortages, diminished farmland, toxic air pollution, horrific crowding and mind-numbing expansion away from nature—environmental refugees will escape, but as the rest of the United

States adds that next 100 million, and then another 100 million, and yet another 100 million—where will anyone make their escape?

"Camp out among the grass and gentians of glacier meadows, in craggy garden nooks full of Nature's darlings. Climb the mountains and get their good tidings. Nature's peace will flow into you as sunshine flows into trees. The winds will blow their own freshness into you, and the storms their energy, while problems will drop off like autumn leaves." John Muir, 1838—1914 ("Yellowstone National Park," Atlantic Monthly, April 1898, 515-6; ONP, 56.)

Will any of what Muir describes be there for future generations?

SECTION II: ENVIRONMENTAL

CHAPTER 4: FOR LACK OF WATER, I'M SO DRY, I CAN'T SPIT

"Water is essential for all dimensions of life. More than eighty countries, with forty percent of the world's population, are already facing water shortages, while in this century the world's population will double. The quality of water in rivers and underground has deteriorated, due to pollution by waste and contaminants from cities, industry and agriculture. Over one billion people lack safe water, and three billion lack sanitation; eighty per cent of infectious diseases are waterborne, killing millions of children each year."

World Bank Institute

We owe our children, and theirs—a sustainable future.

We owe our planet-home reasonable and responsible behavior that complies with the laws of nature. As the most prolific species on earth, we face harsh realities.

First question: what provides the most important aspect of human existence? Answer: clean water!

The latest warning signs manifested at Lake Lanier, Georgia in November 2007. One reporter said to ABC's World News Tonight anchor Charles Gibson, "They need a lot of rain because they're down to the last 36 days of supply for the Atlanta area."

"How much rain?" Gibson shot back.

"Four months of rain would be a good start," the reporter said.

If ever a wake-up call, the vanishing waters of Lake Lanier portend water shortages for five million people—today! Nonetheless, the Peach State expects to grow from 8.2 million people in 2009 to 16.4 million by 2050. Hello! Knock, knock! Anybody home?

On February 21, 2008, anchor Brian Williams at NBC reported that Georgia legislators wanted to extend that state's northern border two miles north in order to annex the Tennessee River. That would allow them to stick a big pipe into a new water source.

As Goober of "Mayberry RFD" might say, "That's like tryin' to milk a cow while sittin' on a stool six feet away. 'Bout the only thing you gonna' get is a tail swishin' full of…well Sheriff Taylor, you know what I mean."

"Yeah, I know what ya' mean Goob," Andy said.

Two thousand miles west of Georgia, Charles Gibson on a February 8, 2008 broadcast, said, "Scientists say Lake Mead, which provides water for millions in the west, expects to go dry by 2023. It's caused by drought, climate change and human population growth."

Colorado State University Professor Neil S. Grigg, on February 17, 2008, in the Denver Post, wrote, "*Not a Drop to Spare.*" He reported, "Colorado's water supplies are nearing their limits, and there is little hope for new sources. What's next?"

To give you an idea of the depth of the West's water crisis, Denver Post journalist John Ingold wrote, "Ship the Mississippi River to Colorado", March 3, 2009. Ingold interviewed Denver businessman Gary Hausler who wants to build a $22.5 billion pipeline to suck water from Old Man River over 1,000 miles from Denver and a mile up hill. Hausler said, "It's saner than any other idea the state's come up with for meeting future water needs."

"Flat-out dumb, stupid, inane, insane and ridiculous beyond comprehension," a reader responded. "What happens when Colorado adds that next five million and then, 10 million and after that, 20 million and more like California? What do these people use for brains…oatmeal?"

None of media reported that Georgia's current population of 8.2 million would double to 16.4 million in four decades. Colorado expects to double from 4.6 to 9.8 million and beyond. California expects to add 20 million in 30 years. Never once did any of the experts pin the needle on the population donkey! It makes you wonder; who made hyper-population a sacred, untouchable cow? Why? Who do they expect to benefit?

Growth and rampant population

As you may appreciate, 'carrying capacity' becomes the most important phrase in our 21st century vocabulary. It entails the amount of human and animal life a limited area of land can sustain in perpetuity.

As this population overload advances, we face major water dilemmas.

"While America remains in her 'consensus trance' brought on by decades of unlimited growth and resources, her citizens cannot imagine water shortages," said James Howard Kunstler, author, *The Long Emergency*. "The vast majority of the earth's surface consists of water, yet only three percent of that is fresh water."

The World Bank famously declared, "The wars of the twenty-first century will be fought over water."

Kunstler continued, "The United Nations identified three hundred zones around the world that will be the sites of conflicts over water in the years ahead. The great aquifers of North America, China and India are all depleting rapidly due to aggressive irrigation…the rapidly diminishing supplies of fresh water, especially in the heavily populated third world, also exacerbate sanitation catastrophes, and prepare the stage for epidemic disease. More than two million people worldwide die every year from contaminated water. In the Maquiladora zones of Mexico today, water is so scarce that babies and children drink Coca-Cola instead."

In a September 30, 2006 Rocky Mountain News report, Boulder, Colorado scientists predicted grim drought forecasts for the West. To support their claim, they used eighteen of the world's most powerful computer climate models. Martin Hoerling of the National Oceanic and Atmospheric Administration said, "Climate change is moving us in the direction of a perpetual state that is of the Dust Bowl type."

Scientists expect increased evaporation and drier soils leading to more severe and frequent droughts. Hoerling said, "Droughts could be 25 percent worse than the 1930s Dust Bowl days."

Who stands to suffer the greatest risk? In 2009, citizens of Utah, Arizona, Nevada and California, downstream of the Colorado River, devour 13.5 million acre feet of the river. That equals millions upon millions of people! Bob Reynolds of NOAA said, "We're going to have to adapt our survival strategies to coping with less water."

Associated Press writer James McPherson on July 30, 2006 wrote a piece for the Denver Post, *"Without Rain, Dakotas Dry Up."* He reported, "Fields of wheat, durum and barley in the Dakotas this summer will never end up as pasta or bread…what is left is hot winds blowing clouds of dirt from dried-out ponds."

More than 60 percent of the United States suffered abnormally dry or drought conditions in the summer of 2006. While I traveled 20,300 miles through 48 states in June, July and August of that year, I saw burned up corn and pigmy crops from lack of water. The drought stretched from Georgia to Arizona and from Montana to Wisconsin.

In the Salt Lake Tribune, January 1, 2009, "The West is hurtling toward a water crisis." Bernard DeVoto said, "The future of the West hinges on whether it can defend itself against itself."

Will America experience commensurate rainfall to provide food and water for that added 100 million? Will we be able to feed and water our 306 million already in the USA? Can science produce miracle crops that grow without water?

Emphatic answer: no!

According to Mike Matz, *"Losing Spaces"*, Denver Post, December 23, 2007, "American farmland and wetlands vanish at 6,000 acres per day, which equals 2.19 million acres annually for new malls, highways and housing. Ground water stores cannot recharge."

Along with lack of water, we degrade water quality. Californians buy more filtered water than anywhere else in America. Why? They can't provide enough clean water to their 37.5 million residents. What about polluted and *'chemicalized'* water run-off? We spray crops, inject insecticides and apply herbicides onto millions of acres of farmland. It seeps into our groundwater and runs into our rivers. The Mississippi River relentlessly spews millions of gallons of fertilizer and chemically poisoned water into the Gulf of Mexico that creates a 10,000 square mile dead zone where few fish or native marine life can survive. Every river running out of the United States, and most industrialized nations, carries enormous amounts of poisons. Acid rain from toxic air pollution falls with every rainstorm.

With blinders securely in place, we pursue rampant population growth with no concern toward future generations. Big surprise—people need water to survive.

The World Health Organization reported in 2006, "Thirty-five percent of humanity doesn't have access to clean drinking water." For a quick reality check, that's more than two billion people. Is America immune to water shortages?

The United States' growing water problems

Short answer: no!

"Is a multibillion-dollar tax hike that could boost water bills as much as 50 percent again hanging over New York City?" said New York Post writer Carl Campanile. "The threat comes from the impact of three large upstate real-estate development projects bordering reservoirs that feed the city's drinking-water supply."

Campanile reported—the state's watershed inspector general, Robert Tierney, raised red flags over these projects:

* A 2,000-acre Catskill Park resort complex surrounding Belleayre Mountain Ski Center could pollute the Ashakon and Pepacton reservoirs.

* A 273-unit housing development and mall in Putnam County could increase pollution in the Croton Reservoir.

* A 104-lot subdivision - also in Putnam - could have a deleterious effect on the Muscott Reservoir.

Pollution threatens drinking water throughout the U.S.

Humans and wildlife stand at risk.

Campanile said, "Pollution, sure to be generated by the developments, likely will prompt the federal Environmental Protection Agency to force New York City to build a water-purification complex costing billions to construct and hundreds of millions annually to operate.

"These costs would be passed directly to property owners. That would mean hugely inflated water-tax bills for single-family homeowners - particularly in residential Queens and Brooklyn."

Campanile continued, "Even worse would be the impact on owners of rental properties. Unlike electricity or gas, there is no way to effectively measure per-apartment water use in a single-meter building. Thus, landlords would be stuck with the full burden of higher water and sewage rates—while being severely restrained by rent regulation from passing the costs along. This could drive marginal properties into bankruptcy."

Former New York Governor Spitzer, who lost his job because of one too many visits for 'dance lessons' at late night bordellos, said, "I'll be neutral while trying to negotiate a compromise between the Catskills' developers and the city's interests."

Overwhelming development misguided

Instead, Spitzer promised to champion economic development there.

Do you see how this kind of thinking drives our civilization over a cliff? In the face of water shortages or damages done to rivers as well as the environment, a sitting governor pressed for more development. Spitzer's successor, Governor David Paterson, follows in the same path. That means more destruction of the wild, which, in turn creates a cascading effect on everything in our environment.

Let's further see how ironic, useless and inept our leaders prove themselves as they pretend to face our future water dilemmas caused by overpopulation.

Leaders lack understanding of population's impact on our fragile environment

Newsweek, April 16, 2007, "*Leadership and Environment*"[5] interviewed Governor Arnold Schwarzenegger, *The Green Giant*, with a picture of him puffing on a cigar. While he filled his lungs with toxic chemicals from the stogie, 37 million Californians' cars, trucks, ships, power plants and homes filled the skies with enormous pollution exhaust.

Schwarzenegger promised to reduce California emissions in 2020 by 174 metric tons.

What didn't he promise? What didn't he address? What did he ignore?

California expects another added 40 million people, if current immigration trends continue, by 2050, to reach a low of 65 million and a high of 79.1 million. (Source: "*US Population Projections for 2050*" Fogel/Martin, March 2006) What does that mean? It means that nothing will be solved. Every aspect of California's accelerating consequences will be multiplied by hyper-population growth.

Not one "Leader" in Newsweek mentioned root-cause

Later in Newsweek's presentation, they featured mayors of cities across America taking the lead. Again, none of their actions will work because none addresses the certain negative impacts of adding 100 million people.

Newsweek's Karen Springen wrote another article on the effects of human population in the same issue, "*Will Polar Bears Be OK?*" Again, Springen failed to mention human overpopulation or anything about stabilizing humanity's numbers!

More people, more combustion engines, more greenhouse gases

Another Newsweek piece, "How to Live a Greener Life" by Jessica Ramirez, presented effective methods for curbing billions of metric tons of greenhouse gases created in the U.S. annually. She prescribed 'powder puff' solutions that looked nice and made people feel good. Such 'Kool-aid' solutions fail miserably. She suggested planting a few trees! As if trees can compete with millions of combustion engines burning 20 million barrels of oil daily! They cannot!

Newsweek continued with a report on China's water crisis caused by its staggering 1.3 billion people. Journalist Orville Schell reported, "The most dramatic national transformation in human history is being threatened by a lack of water. More than 70 percent of China's rivers are severely polluted. One can drive a hundred miles in any direction from Beijing and never cross a healthy river. Many rivers have dried up from human overuse. In 80 percent of rivers still flowing, water quality has been rated 'unfit for human contact' as well as agricultural or industrial use."

Please take note that whatever polluted waters jettison from rivers around the world into our oceans, that contaminated water swirls into every sector of the globe. No marine plants or creatures can withstand growing concentrations of human-made poisons.

China: wall-to-wall people

I've traveled throughout China on my bicycle, up close and personal. China features wall-to-wall people. No let up! No end to it! Accelerating pollution! Compacted living! No escape! Modern Chinese want to imitate Americans—with more resource destruction and added pollution. They create a vicious legacy into the future!

At our current population growth, we follow China and India's disastrous footsteps as we grow toward 1.0 billion people in this century. Fifty years from now, the same report on China's enormous dilemma will become our reality, especially in the West, and throughout the United States.

At least one person in Newsweek's report spoke rationally. K.R. Sridhar said, "I'm not against conservation, but the idea that we can conserve our way out of this problem will not hold."

Mother Nature always wins

Everything reported in Newsweek reminded me of a vision— Governor Arnold, the great muscleman, jumping into the Colorado River north of I-70 in springtime— with a goal of swimming to its source in the Fraser Valley near the Continental Divide in Colorado. The scene proves magnificent! His mighty muscles carry him upward, past the torrents of melting water rushing out of the Rocky Mountains.

Sure enough, the press features him making tremendous progress. Each day, he puffs on his cigar with confidence in his journey. However, the mighty Colorado kept adding water to its annual snowmelt runoff. As Arnold progressed, in reality, rushing water swept him backwards— downstream.

In the end, even Arnold suffered total defeat by the forces of nature.

Connecting the dots-- unchecked, rampant population growth

As you connect the dots in this book, California's fate represents America's future. If we fail to take aggressive steps to address the root-cause, hyper-population growth, we crash, no matter how much we pretend to make progress.

Water? No longer pure! Dangerously polluted! No longer ample!

CHAPTER 5: LOSING THE WILD

"We lose four acres a minute, 6,000 acres a day and 2.19 million acres annually to development caused by population growth in the USA."

Mike Matz, Denver Post "Losing Spaces"

Before the Industrial Revolution, humanity existed by tilling the fields for crops, picking fruits and storing them in root cellars. Transportation included animals, ox carts, rivers and oceans. All limited and slow!

Diseases wiped out millions of people at the drop of a hat. Polio, cholera and bubonic plague ruled.

In 1900, the average American male died by 49 years of age. Citizens kept warm by firewood and coal. As long as we humans depended on solar flow, winds and currents, we remained sustainable within nature's carrying capacity.

However, in the late 1800s, steam power burst upon the scene. With it, steam driven ocean liners and trains afforded swift transport across oceans and continents. With the advent of the internal combustion engine, the tractor and car made their appearance.

Whereas one farmer might feed 10 people with his labors, a tractor allowed one farmer the ability to feed 10,000 humans. Food canning guaranteed sustenance throughout the year.

With the advent of electricity, everything changed in America. Coupled with production and assembly lines, consumption became the driving force of capitalism.

Those technologies allowed Americans to overwhelm the natural world. In 1900, we numbered 76 million in America. At the time, scientists created 100 different chemicals. Today, we surpass 72,000 chemicals with an added 1,000 created annually. All of them outside the bounds of nature! All of them deadly to other life forms including us.

Today the United States, at 306 million people and headed for 400 million by 2035, siphons the lifeblood out of nature at increasing and alarming rates of speed. If we examined the carnage and consumption of our voracious civilization, we might be appalled at the figures we exact on Mother Nature and our fellow creatures.

Each day, Americans slaughter 22 million chickens for consumption. We kill in excess of 105,000 cattle every 24 hours. We devour tens of millions of fish and other ocean life every day. We kill millions of pigs, horses, turkeys, deer, buffalo, ducks, geese, rabbits and other animals. We euthanize eight million cats, dogs and other domestic animals annually.

We burn 7.3 billion barrels of oil annually in the USA. We burn millions of metric tons of natural gas. We burned 1.17 billion tons of coal to produce electricity in 2006.

However, as fast we produce it, we devour it faster. The Sears Tower in Chicago uses more electricity in a single day than the entire city of Rockford, Illinois with 152,000 people.[6] Humans consume 40 percent of the net primary production of energy on earth—the amount of solar energy converted to plant organic matter through photosynthesis—while we make up less than one percent of the animal biomass on this planet.

"It's no accident that as we celebrate the urbanization of the world," said Jeremy Rifkin, president of the Foundation of Economic Trends, "we quickly approach another historic watershed: the disappearance of the wild. Rising population; growing consumption of food, water and building materials; expanding road and rail transport; and urban sprawl—continue encroaching on the remaining wild—pushing it to extinction."

Within the lifetime of our children, vast areas of the wild, that we take for granted, will vanish from our planet. The Trans-Amazon Highway cuts across the entire expanse of the Amazon rain forest, hastening its destruction. What is the result? Harvard biologist E.O. Wilson states that humans create the '*Sixth Extinction Session*' whereby we lose, "Fifty to 150 species a day or between 18,000 and 55,000 species a year. By 2100, two-thirds of Earth's remaining species are likely to become extinct."

Big deal you shrug! As we kill more and more basic plant and animal life, it creates a deadly cascading effect. As humans kill off more and more species, a cascade of extinction destroys environmental equilibrium. Given enough time, we will kill off the grizzly, hummingbird, bald eagle, moose, giraffe, lion, elephant, cheetah, salmon, trout, bass, dragonfly and millions more of earth's creatures.

According to Environmental Magazine, editor Jim Motavalli wrote, "One American uses from 10 to 30 times more resources than a third world person." Thus, our 300 million equates to at least 3.0 billion people using resources. Thus, the next 100 million Americans equal another 1.0 billion humans using resources. Your mind sobers to the accelerating realities we face. Can you imagine adding one thousand cities to the world with 1.0 million residents each in the next 30 years? Name one good reason for that!

Makes your head hurt, doesn't it?

Rifkin said, "In the great era of urbanization, we have shut off the human race from the rest of the natural world in the belief that we could conquer, colonize and utilize the riches of the planet to ensure our autonomy without dire consequences to us and future generations."

Sorry! We cannot get away much longer with our abuse of this planet.

As I've said before and I repeat, we stand, like a proud whitetail buck, in the cross hairs of the most deadly moment in our nation's history. If we fail to stop this Congress from passing pro-growth mass immigration legislation, it will shift into overdrive the greatest importation of humanity ever experienced in the history of the world. It will assure three to four million people added to our country every year. It will not stop illegal immigration; it will explode it. It will not reduce legal immigration; it will double it.

It's as if our citizens by their apathy and our politicians by their ignorance—beg for this country's degradation and collapse.

CHAPTER 6: CROSSING OUR AGRICULTURAL RUBICON

"Colorado lost 1.6 million acres in the 1990s while it grew by 1.3 million people. It expects to lose 3.1 million more acres to concrete and asphalt via development by 2022."

Mike Matz, Denver Post "Losing Spaces"

In 49 B.C., Julius Caesar defied the Roman senate by crossing the Rubicon River to wage civil war against another Roman—Pompey the Great. By crossing the Rubicon, Caesar made a decision whereby he could not turn back.

Today, "Crossing the Rubicon" means: no way to change, repair or undo your destiny. Yes, Caesar conquered Pompey, but the Roman senate, along with Brutus, stabbed Caesar to death. "Et tu Brutus?" Caesar gasped with his last breath.

If the Congress and president sign any kind of an immigration amnesty or double legal immigration in the near future as they attempted in June 2007, they cast the dye; they cross the Rubicon of America's environmental death knell. They most certainly ensure 100 million more people added to our country that explodes our nation to 400 million on our way to a half billion. Once manifested, we will not be able to turn back.

In a crystal clear illustration, "Crossing the Agricultural Rubicon", Dr. John Tanton, spring 2005, The Social Contract Quarterly,[7] presented harsh realities regarding America's food supply.

"We export immense quantities of corn, wheat, soybeans, etc., but much of this crop is fed to animals or processed into food that we then re-import as higher-value agricultural products," Tanton said. "It is the dollar value of imports that is projected to be equal to exports for 2005."

He continued, "The U.S. consumes two-thirds of its own grown food. As population grows, more agricultural land will be converted to non-agricultural uses—roads, hospitals, schools, parking lots, shopping malls and housing projects. Our expanding population will cause us to import more food. The net result will be the gradual decline of our agricultural trade surpluses. We are already in an energy deficit as we import 12 million of the 20 million barrels of oil we burn each day. Now we have a diminishing agricultural exchange surplus with which to buy fuel to facilitate that very agriculture."

The United States feeds the world, but as Tanton exposes in his excellent graphs and charts, we already import as much as we export: "We won't feed people around the world much longer," Tanton said.

For example, Colorado's population will add 1.5 million by 2022. That increase means, according to the Denver Post, those 3.1 million acres of prime farm land suffer development into homes, roads, malls, schools and business parks.

Whatever the population expansion in your state, commensurate farm acreage will be destroyed. For example, by 2050, Texas will grow from 21 million to 48 million people, which means millions of acres of land will be taken out of farming for development. No one knows the disaster that awaits them as to water usage. "Crossing the Rubicon" via farmland destruction brings yours and all states closer to Caesar's fate.

Another aspect of this "Agricultural Rubicon" manifests itself in Eric Schlosser's, *Fast Food Nation,* where he exposes the '*chemicalization*' of our foods by hundreds of additives, colors, preservatives and poisons like the chemical sweetener aspartame.

Since 1950, farmers have sprayed their crops with herbicides and pesticides while injecting soils with dozens of chemical fertilizers that destroy nitrogen fixing bacteria. They poison earthworms, bees and birds, sending them into early graves. Today, we force genetically modified seeds to produce unnatural harvests while we clone many vegetables and create perfect apples. Few customers have bought a 'real' strawberry from a major grocery store chain in the last 20 years. Those genetically manufactured berries are big, fat and white with some red coloring, and taste terrible.

One of my friends upon biting into one said, "This tastes like eating a piece of chalk! This sucks! What happened to real strawberries?" A more sobering reality hits when you appreciate that most kids in our cities don't know the difference because they've never eaten a real strawberry.

The United States Department of Agriculture states that because of depletion of micro-nutrients, you must eat 49 servings of spinach in 2009 to gain the same amount of micro-nutrient value as one serving of spinach in 1949.

In conjunction with fertilizers draining into rivers which poison fish we eat, farm land absorbs acid rain from chemical contaminants raining down from the sky from tens of thousands of industrial smoke stacks spewing sulfur, ammonia, incinerated plastics, mercury and other toxic amalgamations into the air.

In a report, "*U.S. Pesticide Stockpile Under Scrutiny*" by Rita Beamish of the Associated Press, she said, "The Bush administration is seeking world permission to produce thousands of tons of a pesticide that an international treaty banned nearly two years ago, even though U.S. companies already have assembled huge stockpiles of the chemical. Methyl-bromide has been used for decades by farmers to help grow plump, sweet strawberries, robust peppers and other crops, but it also depletes the Earth's protective ozone. The United States and other countries signed a 1987 treaty promising to end its use by 2005."

If you think our government tells the unvarnished truth, think again.

Senator Frank Lautenberg, D-N.J., said he was informed that the Inspector General for the Commerce Department and NASA had begun "coordinated, sweeping investigations of the Bush administration's censorship and suppression" of federal research into global warming. But the total U.S. emissions, now more than seven billion tons a year, are projected to rise 14 percent from 2002 to 2012. In other words, everything that goes up must come down. When it does, it's a disaster for the entire web of life on our planet home.

Lester Brown of Earth Policy Institute notes that farming causes the loss of 24 billion tons of topsoil annually worldwide. Once soil suffers depletion, chemical fertilizers may allow crops to grow, but a consumer may as well be eating cotton candy for the lack of micro-nutrient value in foods.

What about water for irrigation? At the moment, farmers from Iowa to California draw down underground aquifers faster than they can recharge. Farmers suck billions of gallons of water from the great Ogallala Aquifer beneath Nebraska. What happens when it dries up?

Dr. David Pimentel, College of Agriculture and Life Sciences, Cornell University, says if we think growing huge amounts of corn for ethanol fuel provides an option, we need to think that over.

He writes, "Our up-to-date analysis of the 14 energy inputs that typically go into corn production and the nine invested in fermentation and distillation operations confirms that 29 percent more energy (derived from fossil fuels) is required to produce a gallon of corn ethanol than is contained in the ethanol. Ethanol from cellulosic biomass is worse: with current technology, 50 percent more energy is required to produce a gallon than the product can deliver. In any event, biomass ethanol is a bad choice from an energy standpoint.

"The environmental impacts of corn ethanol are enormous. They include severe soil erosion, heavy use of nitrogen fertilizer and pesticides, and a significant contribution to global warming. In addition, each gallon of ethanol requires 1,700 gallons of water (to grow the corn) and produces six to 12 gallons of noxious organic sewage.

"Using food crops, such as corn grain, to produce ethanol also raises major ethical concerns. More than 3.7 billion humans in the world are currently malnourished, so the need for grains and other foods is critical. Growing crops to provide fuel squanders resources. Energy conservation and development of renewable energy sources, such as solar cells and solar-based methanol synthesis, should be given priority."

If we add another 100 million Americans, our impact and consequences multiply by 100 million. That much more chemical spray, fertilizer and water must be used. Remember: for each American added to the United States, 12.6 acres of land must be developed. That's 1.26 billion acres of land expended that can't produce food. Experts tell us that by 2040, we'll be a net importer of food.

What if our food source cannot provide for us? What if we cannot economically transport the food to our shores?

Our country heads into dangerous waters. Have you heard the expression, "Up the creek without a paddle?" Whether it's "Crossing the Rubicon" of agricultural destruction of our food supply, or using up our oil reserves without sufficient alternatives, or exceeding our carrying capacity as to water—we're hyper-populating our nation into grave consequences.

Yet, you won't see the president, his cabinet, Congress or all 50 governors speak about it or address it. Our captains of corporations and industry think they can keep revving the engine of consumption without end.

Most possess a paradigm of economic growth at any cost. Most cannot comprehend their folly as their 'capitalism god of growth' dominates their world view. Most think they can 'red line' the engine of growth by encouraging population without consequences. This behavior resembles that of brain injury victims. They often lack insight, understanding, critical thinking or common sense.

CHAPTER 7: SCIENCE, HUMAN RESISTANCE AND CONSTRAINTS

"How would you describe the difference between modern war and modern industry—between say, bombing and strip mining, or between chemical warfare and chemical manufacturing? The difference seems to be only that in war the victimization of humans is directly intentional and in industry it is "accepted" as a "trade-off." Were the catastrophes of Love Canal, Bhopal, Chernobyl, and the Exxon Valdez episodes of war or of peace? They were in fact, peacetime acts of aggression, intentional to the extent that the risks were known and ignored."

Wendell Berry, *What Are People for?*

As our dilemma accelerates, the picture for overpopulation clarifies for every American. Clint Eastwood as Dirty Harry in "Magnum Force" said, "A man's got to know his limitations."

The same stands for a family, a community, state and country. Everything on this planet exists within limits. A glass of water can only hold as much as it can hold. Only nine players can play on defense on a baseball team on the field at one time. Basketball limits a team to five players. A movie theater holds a limited amount of seating. A plane with a 200 passenger limit must carry exactly that number and not one extra.

David Pimentel, Cornell University, January 4, 2007, said in a speech, "The U.S. population has doubled in the past six decades to over 300 million people. Currently, the U.S. population growth rate is now more than twice that of China. In 100 years, at our current growth rate, the U.S. population is projected to reach 1.2 billion— or nearly the population of China. Is this what we want for future America?

"Like it or not, our natural resources, from land to wood to oil to water, are finite and cannot sustain an infinite population growth without seriously impacting our quality of life. The time has come for government planners and citizens alike to begin weighing the impacts of unabated population growth."

Pimentel added, "More than 99 percent of all our food comes from the land and less than one percent from the oceans and other aquatic ecosystems. Each American consumes more than 2,200 lbs of food per year, and to produce this food requires more than 3.6 acres of agricultural land. Most U.S. cropland is now in production and little is available for expanding food production."

As noted earlier, each added American destroys 12.6 acres[8] of land to support him or her throughout life.

"Along with land, an ample supply of freshwater is essential for food and other human needs," Pimentel said. "Water shortages already exist in many parts of the nation, especially in western and southern states—and such shortages will become more acute if population growth continues unabated. Each American uses about 530,000 gallons of water per year, with about 80 percent used just for food production. For example, an acre of corn requires 500,000 gallons of water during the growing season."

Pimentel added, "More than 90 percent of U.S. oil reserves have already been pumped, and currently more than 63 percent of U.S. oil has to be imported from other nations at a cost of more than $120 billion per year. Yearly, each American uses energy in the equivalent to 2,800 gallons of oil, with 500 gallons devoted just for food production.

"Fossil energy is a non-renewable resource, which means that Americans will require renewable energy sources in the future. Depending on the geographic region, the renewable energy technologies with the greatest potential are photo-voltaics, hydropower, wind energy, biomass (thermal), solar thermal, and passive solar. Yet, even when all solar-based technologies become operational, they are expected to provide only half of the current U.S. energy consumption. These renewable energy technologies will require about 17 percent of U.S. land area for their production—and this is equal to current cropland area in use."

The U.S. produces 4.5 billion gallons of ethanol per year. This uses 18 percent of the U.S. corn crop but the yield represents only one percent of U.S. petroleum use. If 100 percent of U.S. corn were used, the estimated ethanol yield would provide only about six percent of U.S. petroleum needs. As mentioned earlier, ethanol as a viable alternative remains a fantasy.

"The continued expansion of the human population not only is depleting fossil fuels, it is reducing the numbers of native species of plants, animals, and microbes throughout the U.S., many of which are vital to agricultural production processes, such as pollination, and essential for a quality environment," Pimentel said. "Converting land to development and highways takes away valuable cropland acreage. For example, in California 240,000 acres of farmland was lost during last year to development."

Pimentel added, "Highway construction also destroys many thousands of acres of natural habitat for survival of native species. Nearly four million miles of highways cover our land. The area being blacktopped each year is 1.3 million acres (an area equal to the State of Delaware). No species lives under the blacktop. Rapid, unabated population growth, including legal and illegal immigration, stresses our school systems. Some schools have three times the number of students that they can handle with the available teachers and support staff. Overall this lowers our effectiveness of the education system, which in turn reduces the economic viability and competitiveness of the United States in the global market."

Though we feel immune from our accelerating population crisis, it manifests in every sector of our society. "Similarly, the rapid increase in the population is crowding medical facilities in the United States," Pimentel said. "In the past two decades the number of outpatients in hospitals has increased more than two fold, and continues to increase. Some hospitals have been forced to close due to the pressure on their emergency and outpatient facilities."

In California, that number exceeds 86 hospitals and ER wards that suffered closing via bankruptcy in the past five years.

"The rapid population increase in the United States is challenging our food production system, the economy in general, and the environment," Pimentel said. "As humans and their diverse activities expand, the sustainability of the natural environment is threatened and diminished for the future. We, as a nation, must come to grips with the harsh reality that our land, energy, food and water are finite.

The quality of life for us, and especially for our children and future generations, is closely linked to the number of people who live in our 50 states." Professor David Pimentel, College of Agriculture and Life Sciences, Cornell University, Ithaca, New York.

Are you connecting the dots?

CHAPTER 8: AIR POLLUTION

"The tiny particulate pollution from cars, power plants and factories does more than clog your lungs. It leads to development of heart disease, according to a BYU researcher. While exposure clearly impacts the lungs, "long-term, chronic exposure to air pollution seems to manifest more in cardiovascular disease than it does in respiratory disease." The link between air pollution and increased deaths has been shown in research by Pope and others. His most recent study, however, shows the biological mechanism by which long-term exposure to tiny-particle pollution can actually lead to ischemic heart disease, which causes heart attacks, as well as irregular heart rhythms, heart failure and cardiac arrest."

Lois M. Collins, "Pollution in the air can cause heart ills"
Deseret Morning News, December 16, 2003

Recent estimates show more than 100 million Americans breathe polluted air in major cities across America. Air pollution increases lung cancer and asthma susceptibility while injecting tiny particles coated with chemicals into human beings' bodies. Pregnant women breathe poisonous air into their fetus' delicate and developing tissue. Many other health consequences cascade from air pollution.

Every day in America (on average):

* 40,000 people miss school or work due to asthma.
* 30,000 people have an asthma attack.
* 5,000 people visit the emergency room due to asthma.
* 1,000 people are admitted to the hospital due to asthma.
* 11 people die from asthma.
* An estimated 20 million Americans suffer from asthma (1 in 15 Americans), and 50 percent of asthma cases are "allergic-asthma." The prevalence of asthma has been increasing since the early 1980s across all age, sex and racial groups.
* Asthma is the most common chronic condition among children.
* Annual cost of asthma is estimated to be $18 billion.

    *   400,000 Americans die of lung cancer annually.
        Data source: Center for Disease Control, Atlanta, GA 2008

If you live in Los Angeles, Denver, Chicago, Houston and other large cities, you can 'see' the air you breathe. It's brown, yellow or tan in color. It shifts in layers over the skyline. Tall smoke-stacks belch unending streams of poisons from power plants—diesel trucks spew toxic smoke ribbons along the expressways—cars by the millions emit tons of particulates into the air. Millions of homes burn wood, oil and natural gas that exhausts into air over our cities. Sewer systems spew toxic air into our once clean environment. Massive bovine herds emit methane gas by the millions of metric tons.

Do you smoke tobacco products? Why? Why not? Whether you do or not, if you live in a large city, you smoke the equivalent of a pack of cancer producing cigarettes every 24 hours. Your children absorb the same toxins with every breath of their young lives. Your health stands at risk with every breath—because you breathe thousands of toxic particles every day.

In cities like Denver, Colorado, the 'red-warning' flag flies scores of days during the October through April period, typically. No one can burn wood in their fireplaces on red flag days!

During the many summer 'temperature-inversion days' in Denver, as you drive toward the city on I-70 out of the mountains, you can see the 'brown soup' that you are about to breathe. When I return home from a weekend in the pristine air of the Rockies, I'm sickened that I'm back to breathing that toxic air with every breath I take.

Can it get worse?

You bet! Denver expects to add two to three million more people in three decades as the rest of the country adds 100 million people by 2035. Some experts tell us those numbers are much higher. The 2008 PEW Report reported 138 million people added by 2050. It means greater air pollution for every city.

In the summer of 2007 in Denver, Colorado, air quality monitors registered 74 micrograms per cubic meter of particulate —the highest ever recorded for dirty air over the metro area.

As Denver's Rocky Mountain News journalist Todd Hartman reported, "Dirty air over the metro area could linger into today, prolonging a stretch of toxic pollution that has prompted warnings even for healthy people. Air monitors have recorded unprecedented levels of particulates—this is gritty air."

Colorado regulator Mike Silverstein, at the state health department said, "We are praying for winds." Winds from the west tend to sweep the pollution toward the east.

What happens when those winds blow air pollution out of our cities? It falls as acid rain on the land, in our lakes and into our rivers. It poisons the earth and water. It creates havoc with nitrogen-fixing-bacteria in our top-soil. It kills entire fish populations in lakes and rivers. It kills trees and native vegetation by changing their soil "ph" balance. It's insidious, deadly and growing worse. And worst of all, it often finds a route to other nations.

At this time, China places a new coal-fired electrical plant on line every two weeks. The resulting air pollution falls on North America in ever increasing amounts of particulate.

"Among industries, electric power generation has a particularly large impact on the natural environment. Power plants are responsible for:

- 64 percent of all emissions of sulfur dioxide (SO2), the leading component of acid rain and fine particulates;
- 40 percent of all man-made emissions of carbon dioxide (CO2), the leading greenhouse gas believed to contribute to global warming;
- 26 percent of all emissions of nitrogen oxides (NOX), a key component of ozone (smog), acid rain, and fine particulates.

"In addition, water pollution, nuclear waste, toxic waste, and impacts on birds and fish can be attributed to various types of power generation." (Source: U.S. Environmental Protection Agency, *Acid Rain Program*, December 23 1999)

What are we doing about air pollution?

As a result of expensive clean-up actions by industry, emissions have been reduced, but awkward and unplanned growth overwhelms clean-up efforts. Lacking an overall strategic growth-plan, pollution pockets arise overnight.

In the next three decades, if we add 100 million people, that means 50 to 60 million more cars, trucks, planes and trains added to the mix. We will also add at least 30 million homes, thousands of schools, malls, firehouses, sewage treatment plants, commercial businesses, power plants and more air pollution generating facilities. That human growth adds millions of miles of roads, cuts down millions of trees, and paves over millions of acres of beautiful open-space lands–some of which give us food we need every day.

Twenty percent of the oxygen we breathe originates from the green plants on the land and 80 percent stems from phytoplankton in the oceans. As we destroy the land and poison our oceans, we annihilate the mechanisms that produce our life-giving oxygen.

How many other physical health consequences do we suffer as human beings living in air polluted cities? How about eye irritations? How about toxic heavy metals and nasty chemicals transported to our brain tissue—carried by the oxygen from our otherwise life-giving lungs. How about birth defects caused by pollution? How about long-term effects on plant and animal life found in and around air polluted cities?

Willy-nilly population growth continues – unchecked

How long until our eyes open to the certainty that we create unfathomable consequences for future generations by our irrational population growth? Even if we mandated zero population growth in the United States today—by shutting down all immigration, our own population momentum would add 35 million people to this country in the generations born here by 2040. In other words, we're placing future generations into a no-win situation.

Our civilization stands frozen on the tracks—ironically bewildered—in the headlights of a runaway population freight train of our own making.

CHAPTER 9: SPECIES EXTINCTION

"The American people today are involved in warfare more deadly than the war in Vietnam, but few of them seem aware of it and even fewer of them are doing anything about it. This is a war that is being waged against the American environment, against our lands, air, and water, which are the basis of that environment."

Norman Cousins (1915-1990)

The 12.6 acres of land needed to support each American—known as the "Ecological Footprint"—removes that land from its natural setting. It must be paved over, planted in crops, built into suburbs, schools, colleges, fire houses, malls, roads and everything that sustains our society. That means the next added 100 million Americans wreak havoc on the natural world in an ever more deadly population dance. We reduce habitat for all other plant and animal life by 1.26 billion lost acres.

With the United States growing by three million people annually, 37.8 million acres degrade from the natural world into the unnatural. That means animal habitat diminishes by 37.8 million acres each year. They can't eat, drink, find shelter or procreate their species. Consider annual road-kill: animals cannot even make it across our highways without being slaughtered! One million vertebrates suffer death **daily** in the United States from crossing our highways. They include eagles, foxes, armadillos, road runners, rabbits, snakes, deer, moose, elk, antelope, coyotes, gophers, song birds and hundreds of other species. That's one animal run over and killed every 11.5 seconds! It equals 365 million creatures killed each year. (Source: High Country News, Paonia, CO, *"Road Kill Statistics"* February 7, 2005)

Within the past 100 years, Americans, ever expanding across the land, destroyed 50 percent of all wetlands in the lower 48 states. Those former water sources no longer exist for ducks, geese, beavers and millions of other animals.

Today, 6,330 animal species in North America teeter on the edge of extinction. The National Academy of Sciences predicts 2,500 plants and animals go extinct every decade from habitat loss. At some point, these extinction rates create a 'cascading effect' for all other dependent species. We all answer to an intertwined ecosystem.

In the West, the prairie dog provides sustenance for 67 other creatures in the food chain. Over half of prairie dog colonies suffered human development destruction in the past 30 years. As their numbers plummeted, every creature depending on those rodents declined commensurately. In Denver, where eagles and hawks once soared daily in the skies, few remain. In their place, you can see, however, a brown pollution cloud from horizon to horizon.

In southern California, those majestic California condors do not soar on heat thermals any longer, but must be kept in cages to preserve a few dozen left alive. Most duck populations suffered from 25 to 50 percent decline in the last half century. The average citizen cannot see what's happening—therefore, no reaction!

You're invited to step into growing 'unseen' consequences pushed by overpopulation. Chief Seattle said, "What humans do to the web of life, they also do unto themselves."

In a report in the Boston Globe, October 19, 2006, by John Donnelly, "*Scientists Alarmed at Loss of Pollinators*," he reported rapidly dropping numbers of birds, bees and bats could impact humanity's food supply. Most plants depend on their pollen being picked up by birds, bats and bees to be distributed so that fruits, nuts and vegetables can reproduce and grow.

In 2007, on Colorado's eastern plains, farmers imported beekeepers with their mobile bee hives to pollinate crops. The lack of bees in the United States created the first imports of bees since 1922. Reports suggest bees being killed by pesticides called *neonicotinoids*, that impair the bees' immune systems. One of the most widely used is *imidacloprid*. It's sprayed all over crops from California to Maine. Consequently, bees get sick and die!

"In addition," Donnelly reported, "wild pollinators from bumblebees to butterflies to nocturnal moths—have lost much of their habitat, due to vast use of pesticides and herbicides that kill plants and hedges in which the insects and birds live."

Most Americans reading this book eat fruits and vegetables from fields sprayed with poisons and from soils injected with chemical fertilizers. Is it any wonder, in the secondary arena of our dining rooms, we ingest 'chemicalized' foods that cause us cancers? Can you imagine what happens to the birds, bees and bats—not to mention other insects—that die or become mutated by man-made chemicals? I lament Americans' total disregard for their fellow creatures.

In 1800, estimates included one billion bison roaming the western prairie. A billion carrier pigeons blackened the migratory skies. When I was a kid, geese and other birds flew over our house in wave after wave, week after week. No more! You're lucky to see a few hundred buffalo in Yellowstone National Park. You'll never see a carrier pigeon because Americans shot them into extinction. Grizzlies regress into tenuous existence. I find it hilarious and ironic that Americans leap out of their cars to photograph moose, buffaloes, grizzlies and other great beasts, but they won't stabilize their own numbers to allow such creatures enough habitats to live.

Einstein said, "There are two infinites: the universe and human stupidity! I am not sure about the first but I am certain about the second."

With our culpability for causing the extinction of 2,500 plants and animals every 10 years, that number can only grow in the coming decades as we encroach further into their habitat. What moral and/or ethical question does that bring to mind? What right does a cognitive species pretend to possess to create such a killing spree on other species that can't think or fight for their survival? How far and how many other North American species do we expect to vanquish forever to satisfy our relentless expansion?

What might be the optimum number of extinct species that would fall short of the "cascading effect?" At what point would we supersede the "cascading effect" to create an avalanche of even more extinctions of other creatures that depended on the web of life?

At what point would that affect human survival as in the case of the pollinators?

As you can see, we already create horrific consequences in the natural world with our current 306 million Americans. It's not only here in the USA, but worldwide!

As you will read in "Chapter 18: Destroying our Oceans," PBS showed hundreds of thousands of tons of discarded fishing nets retrieved by Scuba divers. The nets had been destroying reefs and marine life because nothing in nature could deal with the nylon. It rolled around the ocean bottom, washed by eternal tides, while it destroyed millions of marine creatures caught in its indolent grasp.

Fishing captains cut it loose—knowing the kind of death their nets created for all marine life victimized by those man-made products. On a worldwide scale, we kill 100 million sharks annually along with uncounted other creatures. How morally unconscionable and utterly reprehensible!

Let's fast-forward to 2035 with another 100 million people added to North America. Remember, the human race globally will have added two billion more humans by that time. Their impact can only multiply our global impact for devastating species extinction unprecedented in history. In fact, scientists tell us that five extinction sessions occurred since the dawn of time. The sixth one moves forward in this century. What causes it? The human race!

At some point, nature resembles a house made of cards—delicate. Humans resemble Katrina's destructive power in a Sri Lanka tsunami-type process. How far down that gopher hole can we afford to go and how will it affect our children at mid century?

SECTION III: POLITICAL AND CULTURAL RAMIFICATIONS

CHAPTER 10: PILING UP ON THE ROCKS

"Unlike plagues of the dark ages or contemporary diseases (which) we do not yet understand, the modern plague of overpopulation is solvable by means we have discovered and with resources we possess. What is lacking is not sufficient knowledge of the solution, but universal consciousness of the gravity of the problem and the education of the billions who are its victims."

Dr. Martin Luther King, Jr.

This book may sober countless readers. It may depress many and cause despair for the faint of heart. However, most Americans can't or refuse to grasp our dilemma. By checking our growth rates, if we don't change course, not only will we add 100 million in three decades, we'll add another 100 million on top of that, and do it again until we become one billion people by the start of the next century.

Few want to discuss it. Everyone 'hush-hushes' about the pregnant preacher's daughter. No one enjoys addressing sexual or domestic abuse now epidemic in America. Few discuss 18 teenage suicides every day in this country. Better not talk about the 17,000 deaths by drunken drivers annually! Let's pretend it's not happening.

"Andy," Barney said, "did you hear about Mary Lou gettin' knocked…"

"Now Barn," Andy said. "Ain't none of our business to go 'round spreadin' rumors."

"Somebody should have given her some advice," Barney said.

Fifty years ago, somebody should have given China and India's leaders some advice, too! They avoided discussion about their exploding populations. Guess what? They suffer today what they didn't address. Imagine what it means to billions of Chinese and Indians now living within the clutches of hyper-population growth.

Adding 100 million people to the United States in three decades might not be so bad if there weren't 300 million already here. But they are here! And, if nothing changes, what do you think newspapers will feature in 2035? Better news than today?

Perhaps you'll read flowery reports on human progress. How about glowing editorials on our expanding American Dream? On the contrary, you'll read, "Water wars pit farmers against city folks for diminishing supplies"; "Gas prices hit $12 a gallon"; "Rolling blackouts can't save pipes from freezing in New York City"; "Food costs soaring due to transportation and production overheads"; "In Los Angeles today, a 100 vehicle pileup caused the deaths of 40, dozens of injuries and created a 50 mile long gridlock on I-10"; "Riots in area high school caused by students' inability to communicate with one another due to language differences"; "Tuberculosis continues its climb as the worst disease outbreak in decades in the United States"; "Millions of Americans moving to Montana, Idaho, North and South Dakota to escape overcrowding in Los Angeles, New York and Houston."

A sobering report in the New York Times, September 30, 2006, said, "India with 1.1 billion people is running through its ground water so fast that scarcity could threaten whole regions. India has 19 million wells, some of them tapping deposits formed at the time of the dinosaurs."

Some 3.3 billion people live in countries that over-pump their aquifers, which includes our own nation.

Mike Davis wrote in his book, *Planet of Slums,* that urbanization of world poverty boils down to this, "Instead of cities of light soaring toward heaven, much of the 21st century urban world squats in squalor, surrounded by pollution, excrement and decay."

Michael Specter, *"The Last Drop,"* The New Yorker, October 23, 2006, said, "Nearly half the people in the world don't have the kind of clean water and sanitation services that were available two thousand years ago to the citizens of ancient Rome. More than a billion people lack access to drinking water, and at least that many have never seen a toilet. In the past decade, more people have died from diarrhea than people have been killed in all the armed conflicts since World War II."

China, at 1.3 billion people, experiences mind-numbing environmental and social problems. The Wall Street Journal headlined *"A Poison Spreads Amid China's Boom"* September 30, 2006, "Toxic sludge sinks villages and people die without recourse. The lack of pollution controls has contaminated China's soil, water and air with lead, mercury and other pollutants and left millions of children with dangerously high levels of toxic metals in their blood."

Every consequence experienced by India, China and Bangladesh stems from hyper-population growth. Every aspect of their human suffering stems from too many people. Every condition heaped upon their citizens stems from disregard of a rational and sustainable population policy. All the while, the elites live above it while the people sink deeper into its clutches.

As William B. Dickinson,[9] author of the, *Bio-Centric Imperative,* said, "Our cavalier attitude toward big population increases never ceases to amaze. When the U.S. hit 300 million in October, the New York Times concluded in an editorial, October 11, 2006, that, 'In America, growth and vitality are the same thing…our population issues have mysterious ways of working themselves out.'"

That's like saying if we can put 10 basketball players onto one court to play a game, we surely could put 100 on the court and we're sure everything will mysteriously work out. Or, let's cram 10 people into a VW bug for a 100 mile trip and expect everything to mysteriously work out. Or, well, you get the idea!

If the New York Times expresses that kind of balderdash with their statement on "mysterious ways of working themselves out," we might as well return to the Dark Ages where reason and critical thinking suffered under religious dogma.

E.O. Wilson wrote: *"The Creation: An Appeal to Save Life on Earth."* He describes humanity as "The giant meteorite of our time…a species blinded by ignorance and self-absorption."

Which brings me to the question: are Americans as thoughtless and docile as the Chinese and Indians, or citizens of Bangladesh fifty years ago? How far down the population gopher hole do we allow our nation to scurry? At what point on the demographic misery scale will we begin to say, "Enough is enough!" How far into the negative headlines, "water shortages, climate change and gridlock" do we want our children to suffer from our folly?

I still don't think many in America 'get it' as to our perilous path. Oprah, 60 Minutes, Charlie Rose, Night Line, Larry King Live, Date Line, Good Morning America, Today and The Early Show have yet to feature any national spokespersons on this issue. Hannity, Colmes, Situation Room, Cavuto, Mathews, Carlson, Face the Nation, Meet the Press and dozens of other programs reject the topic outright. No matter how many graphs, facts, figures and information sheets sent to top newspaper publishers across the country—no response! They choose 'silent assertion' over critical thinking.

In three decades, it will be too late to wake up. Too late to change course! Too late to take action!

Our water crisis will be multiplied by hyper-population. Our gridlocked traffic and crowding multiplied by endless cars. What is it that few understand about this vast, accelerating 'American dilemma' that already plays out in much of the world?

As hyper-population growth adds to America, you and/or your offspring will slog through the consequential muck created by that added human deluge.

"Andy," Barney said, "do you think it's gettin' too crowded in Mayberry with them two new families movin' in?"

"I 'spect we might have to move out to the country to git away from all them city folks," Andy said.

"Looks like everything's headin' into the chicken coop," Barney lamented.

"Can't argue with ya' Barn," Andy said.

Other voices share Barney's concern. Canadian population activist Tim Murray said that Canada, much like America, places everyone in a surreal nightmare, "We are in a lifeboat paddling frantically away from a passenger ship that sank with too many people and too few boats, hoping the suction of the disappearing vessel doesn't draw us under. Of greater danger though, are the hundreds of freezing unfortunates in the water who are swimming toward us. If they all clamber aboard, our boat will capsize and we'll all suffer hypothermia and drown. Our lifeboat has a limited carrying capacity of 20, but the economist on board thinks that we can help an unlimited number of swimmers. With more aboard, we will 'grow' the boat. The clergyman and refugee-advocate passengers argue it is more important to rescue human-rights, abuse-victims than to think about the boat's carrying capacity. The multiculturalist wants more diversity in the boat, and the politician wants to court more potential ethnic voters."

In the end, whether Canada's population crisis or the United States' hyper-population growth dilemma, both countries cannot keep adding people to their lifeboat. If they do, everyone goes down.

CHAPTER 11: POVERTY IN AMERICA

"Can you think of any problem in any area of human endeavor on any scale, from the microscopic to global, whose long-term solution is in any demonstrable way aided, assisted, or advanced by further increases in population, locally, nationally, or globally?"

Dr. Albert Bartlett

While attending a conference in Denver, Colorado, I noticed dozens of beggars on the streets. They stood at intersections carrying cardboard signs that read, "Homeless, anything will help, God bless."

As I walked out of the Colorado Convention Center, one man, wrapped in rags, curled himself around a steam vent on the sidewalk. A cardboard box served as a pillow. Cement served as his mattress. I sickened at the thought of his night in 15 degree freezing cold. Along Colfax Avenue, hundreds of homeless begged for food or money. They slept in huddled misery under the loading docks and in doorways.

The poor will always be with us; just more of them!

The National Coalition for the Homeless, www.nationalhomeless.org reported 3.5 million homeless people struggled for survival in the streets of America in 2006. Of that number, 1.35 million consist of homeless children. Reports show 13 million American children suffer daily from malnutrition and hunger in America. A shocking 37 million Americans live below the poverty line, which is 12.7 percent of our population.

Educational experts estimate over 30 million people in America suffer functional illiteracy. They cannot read, write or perform simple math. They offer no skills other than the labor of their hands.

Twenty million illegal aliens residing in America make up the largest high school dropout population in the history of the nation.

Over half of all black and Hispanic babies originate from unmarried mothers that lack high school diplomas. Fifty to 70 percent of blacks and Hispanics do not graduate from high school. Thirty percent of whites do not graduate from high school.

In the last century, Mexico grew from 50 to 104 million people. Current demographic figures show Mexico growing to 153 million by mid century. Since 85 percent of all immigration into the United States originates from Mexico, we face a striking dilemma.

Can we incorporate a massive and growing illiterate population? How will we contend with the next added 100 million Americans featuring scant educational skills? How will we deal with millions of babies from their ranks? If we cannot educate half of current American minorities, how will we educate this massive overload of humanity?

If Mexico cannot maintain a successful society with 104 million people, how do you suppose they will sustain their country with 50 million more?

To give you a harsh view of our future, I offer an eye-witness account from my bicycle seat traveling through Mexico. On the outskirts of Mexico City with 22 million people, in excess of two million people live in cardboard shacks. They squat for their morning constitutional with their chickens. They live in abject misery, filth, disease and hopelessness.

Guess what? They're moving to America. Millions of them!

Third world slums began appearing along our borders from Brownsville, Texas to San Diego, California in the 80s. Called "Colonias," which in Spanish means "new neighborhoods," they feature shacks, no sewers, no streets, no running water, no electricity, toilet facilities or waste pickup.

The New York Times, March 3, 1988, *"Along the US Border, a Third World is Reborn,"* reported, "Colonias are rusted trailers and shacks nailed together from tar paper and packing pallets without indoor toilets...with mounds of uncollected trash that attract rats... the lack of sanitation has polluted the ground water to the point where many residents drink their own waste...the colonias feature third world levels of hepatitis, dysentery, diarrhea, skin rashes, cholera and tuberculosis...they are contaminated, explosive, fecal, filthy, illegal, miserable, polluted, powder kegs, putrid, shocking, sick, stench filled, suffering and wrenching."

Since their appearance in the early 80s, according to the Times, the 1988 population totaled 185,000; the 1995 population exceeded 500,000; the 2005 population exceeded 1.5 million. At the current rate of growth, the New York Times predicted those human misery settlements would reach 20 million by 2021.

While filming colonias in Texas, I haven't been as sickened to my stomach since my travels in Asia or countries like Haiti. It's worse than any description the New York Times or I could give you. Colonias represent human misery at its disturbing worst levels.

These slums represent a health hazard of unprecedented dimensions. Given enough time, large areas of southern California will resemble the outskirts of Mexico City. Two decades of denial continues the expansion of American colonias.

We cannot import millions of desperately poor, illiterate and hard working people from third world countries and think they will become functioning, positive aspects in a first world country. France, Holland and England's failed immigration policies offer proof. Ours fails, too!

What about our working poor? How about degraded educational opportunities for our children?

As we choke on millions of people from other countries, they displace our working poor as immigrants depress wages. Our educational systems sink in academic excellence, creating millions of added poor.

As it stands today, millions of Americans can't pay for heating and electricity bills. They rely on donations by other Americans to cover those bills. At some point, as this new poor class expands into millions upon millions—something else will fail. What is that? Our ability to deal with it or solve it!

Anyone with an ounce of common sense or economic intelligence knows that prices in the coming years will rise as oil becomes more expensive. This translates into diesel that drives trains that bring coal to the electrical plants. Thus energy at every level will become more expensive. The caveat enters the picture as these millions of poor cannot and will not be able to command higher wages.

The American Dream degrades into the American Nightmare

On the world stage, 57 million people died in 2005. According to Time Magazine, eight million starved to death. Of the total number of deaths, 10.5 million children under the age of 12 years old died from starvation and related diseases.

To bring it into sharper focus, current world population at 6.7 billion will hit between 9.2 and 9.8 billion at mid century. That's 77 million people added annually. They multiply so fast, they cannot be educated. However, they flood into first world countries.

No one, I repeat, no world leader addresses this "human dilemma." The Catholic Church won't allow or talk about life-enabling birth control—though it correlates to accelerating poverty. Church leaders of all the major religions deny any problem. It's almost as if, in the 21st century, they prefer remaining in the 1st century. But, via their actions, millions of adults and children starve to death annually.

We better deal with it: today! To continue on our current path proves as inept as the captain of the Titanic.

CHAPTER 12: FRACTURING AMERICA

"People, I just want to say, you know, can we all get along? Can we get along? It's just not right. It's not going to change anything. Let's try to work it out."

Rodney King, 1992 LA race riots

Today, California sloshes nostril-deep in its 37.5 million population overload. It grows by 1,700 people daily![10] Do the math! By 2050, if this immigration-driven population phenomenon continues, California explodes to 79.1 million people. (Source: *"Projecting the U.S. Population to 2050"* by Jack Martin and Stanley Fogel, March 2006.)

If you think Los Angeles' smog, traffic, water and power shortages create problems today, hold on to your girdle because you ain't seen nothin' yet!

In Dr. Otis Graham's book, *Unguarded Gates: A History of America's Immigration Crisis*, he writes, "Most Western elites continue urging the wealthy West not to stem the migrant tide, but to absorb our global brothers and sisters until their horrid ordeal has been endured and shared by all—ten billion humans packed onto an ecologically devastated planet."

In literary terms, it's called 'hubris' or false pride. Many of Shakespeare's protagonists suffered this self-destructive trait. In the Bush years, his cabinet and Congress epitomized 'hubris' through their incompetent, if not arrogant actions.

How completely unconscious were members of the U.S. Senate when they attempted to pass S.B. 1639 in 2007? It jumped current legal immigration from 1.2 million annually to 2.4 million. That's like seeing a child with his clothes on fire running out of a burning building, but instead of spraying him with water, the firemen spray gasoline on the child. In this case, the current president and Congress pour more 'gasoline' onto the fire by adding millions of immigrants into our country. Note that Obama voted to give amnesty to 20 million illegal migrants and double annual legal immigration.

Scant rational thought or logical actions go into what our leaders force into this nation via immigration.

What are some of the results? Out of the next 100 million added to America, 70 million will be immigrants. When we add that many people from foreign countries to the United States in three decades—we gang-tackle our civilization by adding incompatible cultures, religions and languages. We cripple our educational system. We overload our infrastructure. We demean citizenship. Assimilation cannot occur. We fracture our identity. For a simple example: France burned in December 2005 when immigrants firebombed 10,000 cars, burned houses, killed French citizens and caused three weeks of mayhem.[11]

With the child on fire illustration, what burns in this country? For starters, we scorch the foundation of our society by destroying our most precious bond—our English language. Without a single language, we shall cease to continue as a cohesive civilization.

Samuel Huntington wrote the best seller: *Who Are We*? He stated, "By 2001 Congress appropriated $446 million for bilingual programs, supplemented by huge amounts of state funding." Ironically, so many legal and illegal immigrants find they don't have to speak English because none of their friends speak it. Koreans, Vietnamese, Hmongs, Middle Easterners, Russians and dozens of other immigrants cannot speak English and will not invest in becoming Americans.

Thus, these new immigrants live in America, but they cannot speak, and do not think like Americans. What's more, they could care less. No investment! They enclave themselves into separate groups. Los Angeles stands as a polyglot of seething anti-American bias by Mexicans and their children. One look at Detroit's growing Muslim population portends what happened to Paris, France, i.e., complete separation from the host country by new immigrants.

When you add millions of immigrants that cannot speak our national language, they emotionally, physically and intellectually separate. A nation cannot long survive when its citizens cannot communicate with one another.

As we spin ourselves into massive population overload from third world countries, the labor pool generates millions of unskilled bodies that depress wages. Dr. Vernon Briggs of Cornell University said, "Efforts in the United States to reduce the incidence of poverty have been hampered, since 1965, by the parallel revival from out of the nation's distant past of the phenomenon of mass immigration.

"Unless comprehensive immigration reforms are added to the arsenal of anti-poverty policies, efforts to reduce poverty in the country in the twenty-first century will be little more than a ride on a squirrel wheel—a lot of effort expended but little progress achieved."

As these people avalanche onto our shores, can't speak English and possess few skills, they and we fragment faster than a grenade. They succumb to a profound racism and distrust against their host country.

"Immigrants devoted to their own cultures and religions are not influenced by the secular politically correct façade that dominates academia, news-media, entertainment, education, religious and political thinking today," said James Walsh, former Associate General Counsel of the United States Immigration and Naturalization Service. "They claim the right not to assimilate, and the day is coming when the question will be how can the United States regulate the defiantly unassimilated cultures, religions and mores of foreign lands? Such immigrants say their traditions trump the U.S. legal system. Balkanization of the United States has begun."

For example, Muslims practice female genital mutilation (FGM). It's 6th century Dark Ages barbarism alive in the 21st century. It occurs today in Detroit, Michigan, Freemont, California and Denver, Colorado, and wherever Islam grows in America. Seven years ago, hundreds of cases of little girls suffering from infected genitalia, after being slashed with razors and glass, landed in Colorado hospitals from the FGM ritual. Islamic practitioners, without sterile room technique or anesthesia, cut out the labia majora and minora as well as the clitoris of girls, usually less than nine years of age.

Led by Colorado House Representative Dorothy Rupert, lawmakers passed a new law to halt it. The practice didn't stop; it torpedoed underground. What happens when Muslim culture becomes the majority and votes it into legal practice? What happens when Mexican culture citizens vote horse tripping, cock fighting, bull fighting, Santeria (animal sacrifice already occurring in Florida) and dog fighting into law? What other inhumane practices do other cultures import into America?

Have you seen a country devolve? In the Middle East today, Muslims stone women to death for adultery. They won't allow women to drive. Women cannot go out in public without a male relative at their side. They demand separate swimming pool times for men and women. They make women hide their faces with a burka into 'non-beingness'. They practice Sharia Law which proves harsh and diametrically opposed to parliamentary law.

Don't think it can happen in America? Fox News, July 25, 2008 reported a Muslim honor killing in Garrett, Texas when a 12 year old girl called 911, screaming, "My dad shot me; I am dying." She died before an ambulance arrived. Reason: father didn't like her wearing western jeans and blouses.

In Clayton County, Georgia, July 10, 2008, a Pakistani immigrant father, Chaudry Rashid strangled his daughter for not accepting his choice of a husband. She wanted a divorce. She was 14 when she died.

In New York, February 16, 2009, FOX News, Joshua Rhett Miller reported, "The estranged wife of a Muslim television executive feared for her life after filing for divorce last month from her abusive husband," her attorney said — and was found beheaded Thursday in his upstate New York television studio. Aasiya Z. Hassan, 37, was found dead on Thursday at the offices of Bridges TV in Orchard Park, N.Y., near Buffalo. Her husband, Muzzammil Hassan, 44, has reportedly been charged with second-degree murder. "She was very much aware of the potential ramification of her filing for divorce might have," said attorney Elizabeth DiPirro, whose law firm represented Aasiya Hassan in the divorce proceeding. "But she wanted to proceed despite the potential for it to erupt."

Under Sharia Law in the Middle East, such an 'honor killing' remains accepted in Islamic society. Ann Curry on NBC hosted a news piece, *"Honor Killings in America."* She reported on the astounding rise of honor killings in America by Muslim immigrants.

The new House of Representatives Muslim Keith Ellison (D-MN) took the oath of office January 20, 2007. He placed his hand and allegiance on the Koran. The irony of this stems from the fact that the Koran stands in direct conflict with the U.S. Constitution. Minnesota voters elected a man whose prime directive from his Koran is to convert or kill all non-believers.[12] Specifically, that's all Jews and Christians. He exemplifies the beginning salvo in the disintegration of America.

How far do we want to regress into that arcane religiosity in America? One look at the sectarian violence by the Sunnis, Kurds and Shiites in Iraq should give you pause for concern. One look at the unending violence between Israel and Palestine should give you a clue as to our fate. If you didn't grasp what happened in Paris, France, you will experience it when it erupts here.

As to the 'lifeboat' metaphor for America, how many people does it take to sink a lifeboat? Short answer: the last one to come aboard that overwhelms the craft's ability to remain afloat. When you look around, countries like China, Congo, India, Mexico, Bangladesh, Darfur, Sudan, Somalia and dozens of others already exceed lifeboat capacity.

That brings us, fellow Americans and global citizens, to the yet unresolved decision regarding adding 100 million more people to the United States by 2035. As we degrade our ability to maintain our civilization by adding disparate cultures, languages and the least educated and poorest of the planet, we reduce ourselves to the same kind of conditions they fled. At the same time, not one, single core problem has been solved in their countries. We import their dilemmas into our frayed society.

If you become emotional about this, that won't solve their problems or ours. Bring your greatest sympathy to their plight, but it won't change the fact of *carrying capacity* limits in our civilization. Even if you send food through church group outreach programs, this exacerbates the problem because that food allows third world people to multiply their numbers—causing greater starvation down the road. "More population growth leads to more demand for food," said author of *Peak Everything*, Richard Heinberg. "We are describing a classic self-reinforcing feedback loop."

Such a paradigm proves increasingly lethal. Another 100 million people careens our country into the same consequences as their countries.

This doesn't have to happen! We can change course by our actions. Whatever your passion, you will be able to take action with the last three 'solution chapters' in this book.

"Noble intentions are a poor cause for stupid actions. Man is the only species that calls some suicidal actions 'noble'. The rest of creation knows better." Garret Hardin, author of *Stalking the Wild Taboo.*

CHAPTER 13: LOSS OF FREEDOM

"The wave of the future is not the conquest of the world by a single dogmatic creed but the liberation of the diverse energies of free nations and free men."

John F. Kennedy

In his book, *The Fate of Nations,* Paul Colinvaux wrote, "Liberty, in the Jeffersonian sense, cannot survive a continual packing-in of people. If our numbers continue to rise on a resource base that expands but little, the future inevitably holds ever greater restrictions on individual freedom."

Once again, common sense rears its gut-check propensity as to the dilemma we face with our hyper-population onslaught.

In reality, we've all faced crowded situations where our freedom suffered limitations. Do you recall women standing in line for the toilet at a movie or ball game? How about standing 20 back in line at the movie box office with only 60 seconds before the show? What about a dance floor so crowded you couldn't move? Ever drive in bumper to bumper traffic with a crash scene ahead? How about visiting Yosemite National Park in the summer? Ever try to board a subway car at 5:15 p.m. in New York City? Or, how about walking down the sidewalk at rush hour in NYC? With greater numbers—everyone's freedom suffers limitations.

As your freedoms diminish, your connection to natural processes degrades. At the same time, your stress levels rise not to mention blood pressure and irritation.

As shown in many other historical illustrations, America's fate can be wrecked by wrong decisions concerning population numbers. In Colinvaux's book, he shows that beliefs such as "go forth and multiply without limits" create consequences counter to long term survival of humanity. He writes:

"Civilizations arise from the technical competence of the founders; rising numbers are merely the consequences of that competence as the civilization is able to feed more and more people."

- Crowding impacts the well-off, who invest heavily in child-rearing, more than the poor, who have little to spend on children.
- Trade, often seen as necessary to cope with growth, is not a solution at all; in reality it is a primary cause of growth, robbing nations and people of self-reliance.
- Repression is the elites' means of preventing the middle class from sharing in the benefits largely created by the middle class.
- Revolutions arise from dissatisfied middle classes determined to seize their rightful due.

Most Americans have read about deer populations 'crashing' from too many animals without enough grassland to feed them. As deer begin to starve, they lose their health, freedom and choices. However, as animals, they die from nature's consequences brought about by overloading the *carrying capacity* of the land.

What is the difference for humans? Before farming and the Industrial Revolution, we couldn't produce enough food to explode our populations. In 1850, the world population didn't add up to one billion people in 5,000 years of humanity. Within 150 years of the arrival of the tractor, we've reached 6.7 billion. We are much too clever for our own long-term survival.

Colinvaux hammers readers' minds with, "If social policy promotes unconstrained growth of the poor classes—whether by natural increase or immigration—the effect will be to squeeze out those whose ability created the standard of living that benefited all."

He shows that ecology's first law might be written: "All poverty is caused by continued population growth. Yet a noisy propaganda denies that rising populations cause poverty. We are told by most eminent politicians and international experts that the rising numbers, far from being a cause of poverty, are in fact a result of poverty."

Colinvaux continues, "Three kinds of assumptions have been made in recent years suggesting that our population growth will stop by itself from some inner dynamic of its own. These assertions have had very wide publicity, for they seem to say that the population problem will go away on its own. The assertions stand in grave error."

What are those assertions?

- Populations stop growing as people become wealthy.
- Recent explosive growth in world population has been due to medical advances and will go away as people adapt.
- That human population is now at the inflection point, at which numbers will level off as in other kinds of animals, remaining stationary thereafter.

He said, "All three assertions violate scientific principles and assume that magic is at work in the control of numbers of all livings things."

It matters little whether you remain silent as to political correctness or quiet because you don't want to get involved or feel that it will work itself out on its own. The fact remains: humanity stepped out of the circle of nature at some point in its progress of becoming cognitive beings. We've sidestepped diseases, carrying capacity and even wars as we've grown to 6.7 billion.

However, serious natural consequences await us. Those penalties already manifest for much of humanity as you've seen in this book. Yet, we persist in our arrogance that we can defy nature.

We cannot! That will become more apparent and more realized in the next 10, 20 and 30 years. Why?

As our numbers increase, our freedoms decrease in every aspect of our lives. Where your commute time used to be 15 minutes, it's now 30. Soon, it will be 45 minutes. In California, some employees drive 90 minutes to work and 90 minutes back home. You can think of a dozen more examples—especially if you live in Los Angeles, Atlanta, Houston or Chicago.

Our rising numbers will boil beyond our ability to stop the population caldron from erupting like Mount Saint Helen. We can deny it, ignore it, suppress it and even pretend to fool ourselves for a few more decades—at the most. But the cold hard facts remain! We're already in so much trouble; we may not survive this population 9/11.

At some point, nature will deal with us rather harshly, and with unfathomable methods. As we fight for resources, beg for water, pray for less traffic and all other consequences of too many people, we will wonder how we landed in this billowing, volcanic predicament.

We can and must change direction. We can avoid becoming the Hindenburg, Mount Saint Helen, 9/11 WTC towers or Titanic of our century. As this book winds down, you will be given straightforward solutions to change course. You can do something or nothing. Your actions create the future!

CHAPTER 14: ENERGY AND THE SILENT LIE

"The United States cannot afford to wait for the next energy crisis to marshal its intellectual and industrial resources. Our growing dependence on increasingly scarce Middle Eastern oil is a fool's game—there is no way for the rest of the world to win. Our losses may come suddenly through war, steadily through price increases, agonizingly through developing-nation poverty, relentlessly through climate change—or through all of the above."

James Woolsey, US Director of Central
Intelligence 1993 - 1995

What more will it take Americans to comprehend what it means to add massive population load onto the United States in three decades? How can we rationally and emotionally accept that number of people added in a blink of time?

What is the easiest and most common posture for dealing with our hyper-population growth? Short answer: ignore, deny, discount, pretend, reject or refute!

Another answer pops up! Lie like a thief. Lie like a politician. If you tell the lie long enough and often enough—people accept it as the truth. We discover the lie in the invention of an oxymoron: sustainable growth!

Back in 1860, one of my favorite authors, Mark Twain said, "Almost all lies are acts, and speech has no part in them. I am speaking of the lie of silent assertion; we can tell it without saying a word. For example: It would not be possible for a humane and intelligent person to invent a rational excuse for slavery; yet you will remember that in the early days of emancipation agitation in the North, the agitators got but small help or countenance from anyone. Argue, plead and pray as they might, they could not break the universal stillness that reigned, from the pulpit and press all the way down to the bottom of society—the clammy stillness created and maintained by the lie of silent assertion; the silent assertion that there wasn't anything going on in which humane and intelligent people were interested.

"The universal conspiracy of the silent assertion lie is hard at work always and everywhere, and always in the interest of a stupidity or sham, never in the interest of a thing fine or respectable. It is the most timid and shabbiest of all lies...the silent assertion that nothing is going on which fair and intelligent men and women are aware of and are engaged by their duty to try to stop."[13]

In 2008, a journalist in Colorado sent information packets and graphs to Publisher John Temple of the Rocky Mountain News; Dean Singleton of the Denver Post and 100 other top national newspaper publishers. He sent out similar packets showing our future dilemmas to top ABC, NBC and CBS executives in the Denver area. He informed them how their kids would be victims of 100 million added people. He explained the water crisis. He invited them to interview 10 top national experts to educate the public. He included self-addressed stamped envelopes for them to respond and provided ideas on how to gain national coverage on this population crisis. He also contacted all the producers and directors at 60 Minutes, Prime Time, NPR and Date Line. He sent letters to Brian Williams, Katie Couric and Charles Gibson.

How did they respond? Take a guess! "Silent assertion" to the maximum!

Why? The more papers and advertisements those folks sell, the bigger their Lear Jets and nicer their luxury cars. Don't bother them with consequences!

If it wasn't for the Internet, this country would be sold down the river faster than it is now. U.S. borders would be dissolved and American sovereignty as a nation would be a thing of the past.

The vast majority of Americans don't have a clue as to their future. Furthermore, they trust in those same leaders who use 'silent assertion' to keep them in the dark.

This book informs and educates while offering solutions through actions. You cannot expect Congress to confront this daily accelerating national catastrophe. They represent 'silent assertion' at its finest.

So what is the lie we're talking about?

Stephen Benka, author of "*The Energy Challenge*" outlined the magnitude of our energy crisis by citing projects from the U.S. Department of Energy from 1999 to 2020, "The world's total annual energy consumption will rise 59 percent and the annual carbon dioxide emissions will rise 60 percent while the world population increases from 6.5 billion to 7.5 billion."

Interestingly, few of the top scientists call for population stabilization. Other experts call for—here's that phrase again— unending 'sustainable growth'.

Let's wake up and smell the coff...er, I mean, consumption addiction!

The late Julian Simon, compelling Economics Professor at the University of Maryland at College Park wrote, "Technology exists now to produce in virtually inexhaustible quantities just about all the products made by nature. We have in our hands now the technology to feed, clothe and supply energy to an ever growing population for the next seven billion years."

To that I say Time Editor Richard Stengel and Professor Julian Simon's spirit should climb into a rowboat together to paddle toward LaLa Land. They might bring "Alvin and the Chipmunks" to sing along the way. On their journey, stop by China, India, Mexico and Bangladesh for a dose of reality.

What's that reality? Last winter, Denver, Colorado suffered rolling blackouts because it could not import enough natural gas to heat suburban homes. Concerning natural gas, an expert on gas production in the Clinton administration, Dr. Ernest J. Moniz wrote, "U.S. consumption represents half of that for the industrialized world...with China, Central and South America expecting to triple their usage over the next 20 years."

Where does that leave you and me, average citizens of the United States as our leaders shove this population juggernaut down our throats?

As Colorado University Professor Dr. Albert Bartlett said, "Population growth is given as a cause of the problems identified, but eliminating the cause is not mentioned as a solution. We are prescribing aspirin for cancer."

Bartlett wrote a penetrating column for the local Boulder Daily Camera, Colorado on February 3, 2008, page 7B:

"It's time to try again to correct the educationally credentialed but innumerate experts (innumeracy is the mathematical equivalent of illiteracy) who say that growth is inevitable.

"They fail to recognize that after maturity, continued growth is either obesity or cancer.

"We must remember that "Smart Growth" and "Dumb Growth" destroy the environment in good taste. Growth is not the answer; it's the problem.

"The innumerate theme of the growth promoters is, 'The Front Range of Colorado is going to grow whether we like it or not.' If this is true, it is because so many leaders are active and successful in promoting growth.

"The Legislature and all manner of public and private regional and local civic groups promote economic development which is the politically correct name for growth.

"Predictably, this will produce more well-to-do people, more homeless people, more employed people, more unemployed people, higher salaries, more people living below the poverty level, more traffic congestion, higher parking fees, more school crowding, more crime, more unhappy neighborhoods, more expensive government, more taxes, higher taxes, more fiscal problems for state and local governments, more air and water pollution, higher utility costs, less reliable utility service, less democracy, higher food costs and more destruction to the environment."

"You know Andy," Barney added, "he said a mouthful of common sense."

"You ain't wrong Barn," Andy said. "We got to do something before it gets worse."

"How 'bout we lock up a few politicians to start things off right?" Barney said.

"Can't disagree, Barn," Andy said.

Defining future realities of hyper-population growth

Please consider three definite outcomes by famous economist Kenneth Boulding,[14] who proves a tad sharper than Julian Simon. Boulding offers these three theorems that face our children:

"**The Dismal Theorem**: If the only ultimate check on the growth of populations is misery, then the population will grow until it is miserable enough to stop growth.

"**The Utterly Dismal Theorem**: Any technical improvement can only relieve misery for a while, for so long as misery is the only check on population, the improvement will enable more people to live in misery than before. The final result of improvements, therefore, is to increase the equilibrium population, which is to increase the sum total of human misery.

"**The Moderately Cheerful Form of the Dismal Theorem**: If something else, other than misery and starvation, can be found which will keep a prosperous population in check, the population does not have to grow until it is miserable or starves; it can enjoy stable prosperity."

How far will we sell our national soul for more energy? When will we direct our solutions to population stabilization rather endless growth along with searching for dwindling energy supplies?

Environmental Working Group said, "Want to trade the Grand Canyon for nuclear energy? Your elected representatives didn't, so in June the House Natural Resources Committee passed an emergency resolution to stop the sudden surge of mining claims staked around this treasured resource. But an Environmental Working Group investigation found that in August the former Bush administration ignored Congress and allowed a Phoenix-based mining company called Neutron Energy to stake 20 claims to mine uranium around the Grand Canyon and the Colorado River. Activists have tried to reform our ancient 1872 mining law for decades, and EWG is juicing up the debate with our interactive, updated maps. Stay tuned for more news from the 111th Congress."

Energy and environmental answers from our leaders must no longer be considered apart from a simultaneous decrease of population growth. The very presence of rising numbers forces the search for often terrain destroying solutions.

We don't need to create one more Los Angeles, New York City, Chicago, Atlanta, Denver, San Francisco, Dallas or Houston with big city traffic. We do not need more air pollution nightmares along with millions of miles of highway systems whose diet consists of 20 million barrels of oil daily.

Finally, America reached 'Peak Oil' in 1970. Most predictions show the world oil supplies peaking in 2010. From that point, no matter how many oil wells we sink, the world's supply declines. We live in the 'Peak Oil Descent' that means every oil-dependent civilization on this planet must contract.

James Howard Kunstler said, "At peak oil, there will still be plenty of oil left in the ground—in fact, half of all the oil that ever existed—but it will be the half that is deeper down, harder and costlier to extract, sitting under harsh and remote parts of the world, owned in some cases by people with a grudge against the United States, and this remaining oil will be contested by everyone. At peak and just beyond, there is potential for system failures of all kinds—social, economic and political. Peak is quite literally a tipping point. Beyond peak, things unravel and the center does not hold. Beyond peak, all bets are off about civilization's future."

We need a "consciousness shift" and actions to prepare our civilization for a very different future. You may start at www. transitionus.ning.com.

Whatever you do, avoid becoming a part of the "shabbiest of all lies…silent assertion."

CHAPTER 15: SOCIETAL SELF-DELUSION AND POLITICAL CORRECTNESS

"The politicians are getting so bad around here, these days they'll call a ham sandwich racist!"

Haydee Pavia, Los Angeles, California

Dr. Albert A. Bartlett,[15] University of Colorado at Boulder, presented another rendition concerning America's love affair with self-delusion. Dr. Bartlett remains one of the premier American voices concerning the greatest predicament facing our civilization in the 21st century.

While everyone in America suffers from its symptoms: gridlock, air pollution, crowding and higher energy prices—no politicians from governors to senators address it. Yet, it accelerates as THE single greatest dilemma facing America and the world in the 21st century.

"Throughout the world, scientists are prominently involved in solutions to the major global problems such as global climate change and the growing inadequacy of energy supplies," Bartlett said. "They present their writings in publications ranging from newspapers to scientific journals, but with few rare exceptions, on one point they all replace objectivity with 'political correctness'.

"In their writings the scientists identify the cause of the problems as being growing populations. But their recommendations for solving the problems caused by population growth almost never include the recommendation that we advocate stabilizing our population. Political Correctness dictates that we do not address the current problem of overpopulation in the U.S. and the world.

"We can demonstrate that the Earth is overpopulated by noting the following—*A Self-Evident Truth:*

"If any fraction of the observed global warming can be attributed to the actions of humans, then this, by itself, constitutes clear and compelling evidence that the human population, living as we do, has exceeded the Carrying Capacity of the Earth, a situation that is clearly not sustainable.

"As a consequence it is, *An Inconvenient Truth,* that all proposals or efforts at the local, national or global levels to solve the problems of global warming are serious intellectual frauds if they fail to advocate that we address the fundamental cause of global warming—namely overpopulation.

"We can demonstrate that the U.S. is overpopulated by noting that we now (2008) import 60 percent of the petroleum that we consume, 15 percent of the natural gas that we consume and 20 percent of the food we eat. Because the U.S. population increases by over three million per year, all of these fractions are increasing. Natural gas production in North America has peaked in spite of the drilling of hundreds of new gas wells annually. In a nutshell, the U.S. in 2008 is unsustainable.

"Let's look at two prominent examples of this political correctness. The book, *An Inconvenient Truth,* was published to accompany Al Gore's film by the same name. On page 216 Gore writes; "The fundamental relationship between our civilization and the ecological system of the Earth has been utterly and radically transformed by the powerful convergence of three factors. The first is the population explosion."

"It's clear that Gore understands the role of overpopulation in the genesis of global climate change. The last chapter in the book has the title, "So here's what you personally can do to help solve the climate crisis." The list of 36 things starts with "*Choose Energy-efficient Lighting*" and runs through an inventory of all of the usual suspects without ever calling for us to address overpopulation!

"As a second example, in the Clearinghouse Newsletter (2) we read the statement, "*Human Impacts on Climate*" from the Council of the American Geophysical Union (AGU). The title recognizes the human component of climate change which we note is roughly proportional to the product of the number of people and their average per capita annual resource consumption.

The last paragraph of the AGU statement starts with the sentence, "With climate change, as with ozone depletion, the human footprint on Earth is apparent."

"The rest of the paragraph suggests what must be done, and it's all the standard boilerplate: "Solutions will necessarily involve all aspects of society. Mitigation strategies and adaptation responses will call for collaborations across science, technology, industry, and government." Etc., etc., etc... There is no mention of addressing the overpopulation which the statement recognizes is the cause of the problems.

"A few years ago I wrote an article calling the attention of the physics community to this shortcoming." (3) To my amazement, most of the letters to the editor responding to my article supported the politically correct unscientific point of view." (4), (5)

"Many journalists look to the scientists for advice. The scientists won't talk about overpopulation, so the journalists and the reading public can easily conclude that overpopulation is not a problem. As a result, we have things such as the cover story in Time Magazine, April 9, 2007, *"The Global Warming Survival Guide: 51 Things You Can Do to Make a Difference."* The list contained such useful recommendations as "Build a Skyscraper," (No. 9, Pg. 74) but not one of the 51 recommendations deals with the need to address overpopulation!

"What's one to do when scientists and political leaders demonstrate their understanding of the fact that overpopulation is the main cause of these gigantic global problems (collapsing fisheries, ozone hole, air pollution, acidification of our oceans via human pollution, dead zones in our oceans, species extinction, 18 million humans die of starvation annually, loss of rainforests....), yet the scientists' recommendations for dealing with the problems never call for addressing overpopulation?"

In the past 20 years, I've heard Dr. Bartlett's brilliant lecture on 'exponential growth' many times. He presents his lecture to hundreds of thousands around the country. He spellbinds audiences! His presentation shows where everything looks 'okay' up until the point where a species cannot save itself.

Whether you think humans or natural cycles create climate change, we will not escape this fact: humans grow by 77 million, net gain, annually on a finite planet. They create cause and effect on every environmental system on our planet home—destructively.

As citizens on this blue-green globe out in the black void of space, I encourage you to stand up and make your voice heard during the coming months and years. Since we continue as a fairly stable society at 2.03 children average per American female, we must change course on what causes our demographic Katrina. It's up to us!

(1) Al Gore, *"An Inconvenient Truth, The Planetary Emergency of Global Warming and What We Can Do About It"* Rodale Press, Emmaus, PA, 2006

(2) Teachers Clearinghouse for Science and Society Education Newsletter, winter 2008, Pg. 19

(3) A.A. Bartlett, *"Thoughts on Long-Term Energy Supplies: Scientists and the Silent Lie,"* Physics Today, July 2004, Pgs. 53-55

(4) Letters: Physics Today, November 2004, Pgs. 12-18

(5) Letters: Physics Today, April 2006, Pgs. 12-15

Published in the Teachers Clearinghouse for Science and Society Education Newsletter, Vol. 27, No. 2, spring 2008, Pg. 21.

CHAPTER 16: HOW TO DESTROY AMERICA

"Most Western elites continue urging the wealthy West not to stem the migrant tide, but to absorb our global brothers and sisters until their horrid ordeal has been endured and shared by all--ten billion humans packed onto an ecologically devastated planet."

Dr. Otis Graham, *Unguarded Gates*

Okay, let's get political! Let's cause a ruckus! While this tome addresses our sheer hyper-population load, I felt that a few chapters might deal with the political ramifications. I can think of no greater voice in this arena than former Colorado Governor Richard D. Lamm.

While this nation faces a population overload, the United States must address major cultural differences caused by massive, unrelenting and unending immigration from third world cultures.

On October 3, 2003, I attended a Federation for American Immigration Reform Conference in Washington, DC—filled to capacity with concerned Americans and leaders. Writers, speakers, CEOs, representatives from Congress such as Tom Tancredo as well as former governors graced the podium. Radio talk show host Terry Anderson spoke like a gathering tornado to a capacity audience. Bonnie Eggle, mother of the National Park Ranger Kris Eggle, slain by Mexican drug runners seven years ago on our unguarded southern border—gave a compelling speech that left not one dry eye in the place. Peter Gadiel, father of Jamie Gadiel, spoke powerfully on how the World Trade Center took his son and how nothing has been done since—to stop the flow of illegal migration into the United States.

Even with the façade of Homeland Security, according to Time Magazine, September 12, 2004, "Who Left the Door Open?"—4,000 illegal aliens cross nightly on the Arizona border alone. "Three million continue walking, crawling or tunneling across our borders annually," reported Time. Their accelerating numbers undermine America's ability to function.

During the conference, speaker after speaker astounded the audience with facts on how fast the present administration and Congress continue dismantling the American Dream for average citizens. Mr. Rob Sanchez of Arizona showed how H1 B and L1 visas have ripped one million high tech jobs out of American workers' hands. Another speaker told a packed audience how 'offshoring' and 'outsourcing', fully supported by the president and congress, have cost over three million American jobs in the past six years. His prediction was even more depressing: "In excess of three million more jobs will be 'outsourced' within four years. Those American jobs are headed to Mexico, India, China, Pakistan and Brazil."

Moments later, former Colorado Governor Richard D. Lamm,[16] stood up and gave a speech on *"How to Destroy America."* The audience sat spellbound by the eight methods for destruction of the United States.

Lamm said, "If you believe that America is too smug, too self-satisfied, too rich, then let's destroy America. It is not that hard to do. No nation in history has survived the ravages of time. Arnold Toynbee observed that all great civilizations rise and fall, and that, "An autopsy of history would show that all great nations commit suicide."

"Here is how they destroyed their countries," Lamm said. "First, turn America into a bilingual or multi lingual and bicultural country. History shows that no nation can survive the tension, conflict and antagonism of two or more competing languages and cultures. It is a blessing for an individual to be bilingual; however, it is a curse for a society to be bilingual. The historical scholar Seymour Lipset put it this way, "The histories of bilingual and bicultural societies that do not assimilate are histories of turmoil, tension and tragedy. Canada, Belgium, Malaysia, Lebanon—all face crises of national existence in which minorities press for autonomy, if not independence. Pakistan and Cyprus have divided. Nigeria suppressed an ethnic rebellion. France faces difficulties with Basques, Bretons and Corsicans."

Lamm continued on how to destroy America, "Invent 'multiculturalism' and encourage immigrants to maintain their own culture. I would make it an article of belief that all cultures are equal. That there are no cultural differences! I would make it an article of faith that the Black and Hispanic dropout rates are due to prejudice and discrimination by the majority. Every other explanation is out of bounds."

Any citizen can see that Los Angeles, California or Miami, Florida are no longer American cities. Chicago, Houston, Dallas, Chicago, New York, Atlanta, San Francisco and many other cities are not far behind. Immigrants arrive at over two million per year, so fast and so many, that they do not assimilate into the American Dream. However, they create the American Nightmare for Americans. You would be insulted if you spoke English in immigrant enclaves of those cities.

"We could make the United States another 'Hispanic Quebec' without much effort. The key is to celebrate diversity rather than unity. As Benjamin Schwarz said in the Atlantic Monthly: "The apparent success of our own multi-ethnic and multi-cultural experiment might have been achieved not by tolerance but by hegemony. Without the dominance that once dictated ethnocentrically and what it meant to be an American, we are left with only tolerance and pluralism to hold us together."

Two weeks before the Federation of American Immigration Reform Conference, Muslim students at San Francisco State University charged a student-run Republican voting table. They knocked it over and threatened to blow up the place. One Muslim woman vowed to become a bomber martyr. They did not respect our Republican form of government where each of us is afforded our rights to free speech. Ironically, these were American Muslim student citizens. Apparently, they didn't realize that they lived in America.

Lamm continued, "I would encourage all immigrants to keep their own language and culture. I would replace the melting pot metaphor with the salad bowl metaphor. It is important to ensure that we have various cultural sub-groups living in America reinforcing their differences rather than as Americans, emphasizing their similarities."

"Fourth, I would make our fastest growing demographic group the least educated. I would add a second underclass, unassimilated, undereducated and antagonistic to our population. I would have this second underclass have a 50 percent dropout rate from high school."

In 2005, in Denver, Colorado, 67 percent of the potential graduating class either flunked out or dropped out.[17] The crisis manifested in Denver where Tina Griego, an advocate for illegal immigration and employed by the Rocky Mountain News reported November 15, 2004, "North High School teachers estimate only half the student body of 1,400 regularly attend classes."

That school system suffers hundreds of illegal alien kids who can't speak English and arrive from functionally illiterate parents. They cannot perform academically because their parents have no interest or background in viable educational processes. A similar crisis happens from California to Georgia and from Chicago to Miami.

However, democracy requires four disciplines to flourish while remaining viable. It requires an educated population that possesses a similar moral and ethical foundation while speaking the same language. The United States, via massive immigration, is losing all four aspects. It cannot and will not remain workable much longer with the disruption of its educational systems. Conflicting languages bring a whole new can of worms to the equation.

"My fifth point for destroying America would be to get big foundations and business to give these efforts lots of money. I would invest in ethnic identity, and I would establish the cult of 'Victimology'. I would get all minorities to think their lack of success was the fault of the majority. I would start a grievance industry blaming all minority failure on the majority population."

While a few elites in high places make billions of dollars, they corrupt America's rule of law and undermine the middle class. If you'll notice—third world countries do not sustain a middle class. They suffer a high class and a low class. America proceeds in that direction every day of this immigrant onslaught.

"My sixth plan for America's downfall would include dual citizenship and promote divided loyalties. I would celebrate diversity over unity. I would stress differences rather than similarities. Diverse people worldwide are mostly engaged in hating each other—that is, when they are not killing each other. A diverse, peaceful or stable society is against most historical precedent. People undervalue the unity it takes to keep a nation together. Look at the ancient Greeks: "The Greeks believed that they belonged to the same race; they possessed a common language and literature; and they worshipped the same gods. All Greece took part in the Olympic Games.

"A common enemy, Persia, threatened their liberty. Yet, all these bonds were not strong enough to overcome two factors… local patriotism and geographical conditions that nurtured political divisions. Greece fell. In that historical reality, if we put the emphasis on the 'pluribus' instead of the 'unum', we can balkanize America as surely as Kosovo."

"Before we got here, Miami was nothing," yelled a militant Cuban.

Today in Miami, you will be estranged if you are an American. Point blank reality: Miami no longer represents an American city. Los Angeles, ditto! Every city in America suffers assault. Corruption reigns supreme along with growing "Third World Momentum." Mexico's Calderon dictates his wishes in our country by spreading the fraudulent Matricular Consular ID card while promoting 54 Mexican consulates in major cities.

Lamm continued, "Next to last, I would place all subjects off limits—make it taboo to talk about anything against the cult of 'diversity'. I would find a word similar to 'heretic' in the 16th century— that stopped discussion and paralyzed thinking. Words like 'racist' or 'xenophobe' halt discussion and debate."

At every juncture today, anti-American groups like La Raza, MalDef, LuLac, and Mecha, which promote Mexican groups on U.S. soil—scream 'racism' if anyone protests illegal migration. La Raza's motto states: "For the Hispanic race, everything; anyone outside the race, nothing!"

Think for a long moment about the in-congruency of their message. It does not portend a '*United*' States of America.

You may wonder what will happen when they gain majority populations. Will they allow us to use the word 'racism' as they promote their agenda of 'Reconquista of Aztlan' while they break our country into pieces? With 12-15 million Mexicans now transplanted onto U.S. soil, they are well on their way.

"Having made America a bilingual—bicultural country, having established multiculturalism, having the large foundations fund the doctrine of 'Victimology'," Lamm said, "I would next make it impossible to enforce our immigration laws. I would develop a mantra: "That because immigration has been good for America, it must ALWAYS be good. I would make every individual immigrant sympatric and ignore the cumulative impact of millions of them."

In the last minute of his speech, Governor Lamm wiped his brow. A profound silence swept over the room. Finally, he said, "Lastly, I would censor Victor Davis Hansen's book – *Mexifornia!* His book is dangerous. It exposes the plan to destroy America. If you feel America deserves to be destroyed, don't read that book."

The audience sat stunned. No applause! A chilling apprehension rose like an ominous cloud above every attendee at the conference. Every American in that room knew that everything Lamm enumerated proceeded methodically, quietly, darkly, yet pervasively across the United States as he spoke.

Discussions suffer suppression via political correctness. Over 100 languages rip the foundation of our educational system and national cohesiveness. Barbaric cultures that practice female genital mutilation, honor killings, terror, horse tripping, cock fighting, dog fighting and worse—grow in America as we celebrate 'diversity'. American jobs vanish into the third world as insatiable corporations create a third world in America—take note of California and other states—to date, 20 million illegal aliens and growing, fast.

If you look around you, our past president and possibly this new one, and this Congress dismantle America by design. You may call it 'multi-culturalism' or 'diversity' or 'one world government' or 'globalism'. In the end, it will mean the extinction of the United States of America as a free, cohesive, sovereign and functioning society.

It reminded me of George Orwell's book—*1984*. In that story, three slogans are engraved in the Ministry of Truth building: "War is peace," "Freedom is slavery," and "Ignorance is strength."

It dawned on everyone at the conference that our nation races down a slippery slide toward an uncertain future. An onslaught of diseases, clashing cultures, languages, Balkanization and mounting environmental dilemmas accelerate by the day. If this immigration monster isn't stopped, it will rage across the United States like a California wildfire and destroy everything in its path, especially the American Dream.

SECTION IV: HEALTH DILEMMAS

CHAPTER 17: DEATH, DISEASE & CONSEQUENCES

"Infectious diseases are now spreading geographically much faster than at any time in history. Human immigration and unlimited transport cause it."

World Health Organization

If you travel into the third world such as Mexico, Central and South America, you will notice that while visiting a bathroom you discover a box for used toilet paper in the corner and no soap or paper towels at the lavatory.

The sewage systems cannot handle toilet paper so it is a habit to throw it into the box provided which lures flies and cockroaches. Additionally, few third world people wash their hands after bathroom use. Today, in California, Florida and Illinois, and spreading to other states across the nation, recent arrivals are so accustomed to throwing their used toilet paper into boxes, they discard it into trashcans. Whether they work at the counter or chop tomatoes with unwashed hands, thousands carry head lice, leprosy, tuberculosis and hepatitis A, B, and C.

Annually, an estimated 1.0 to 1.4 million illegal migrants cross America's southern borders—avoiding health screening. Other investigators report larger numbers.

Time Magazine, September 20, 2004, *"Who left the door open?"* Donald L. Barlett and James B. Steele,[18] wrote, "The U.S.'s borders, rather than becoming more secure since 9/11, have grown even more porous. And the trend has accelerated in the past year. It's fair to estimate, based on a Time investigation, that the numbers of illegal aliens flooding into the U.S. this year will total three million—enough to fill 22,000 Boeing 737-700 airliners, or 60 flights every day of the year. It will roughly triple the number of immigrants who will come to the U.S. by legal means."

They fail to stop or be vaccinated for a host of diseases they bring into America. Who is at risk? Everyone, but especially our school children when they come in contact with in-excess of 4.3 million illegal alien school children daily! What can those 4.3 million kids unknowingly transfer to our kids? (Source: www.cis.org)

Tuberculosis kills nearly two million humans worldwide annually. Tuberculosis remained almost non-existent in the United States eight years ago. In 2003, a school in Sebewaing, Michigan[19] reported 30 children and four teachers had tested positive for tuberculosis infections. Michigan supports a large Latin illegal alien population that migrated from Mexico.

In the past eight years, 16,000 cases of TB,[20] which was formerly endemic only to Mexico, crossed over the borders inside the bodies of illegal aliens. Those adults and their children have spread out across the country to work in fast food restaurants and harvesting. Another outbreak occurred in Austin, Minnesota in 2003 where eight police officers tested positive for tuberculosis. After arresting illegal aliens for drunk driving, officers sat in the same car to contract the disease in the vehicle or in the jail where infected illegal aliens waited for trial. A similar outbreak occurred in Portland, Maine with 28 testing positive for tuberculosis.[21]

The Santa Barbara News-Press, April 25, 2004, reported *"Anatomy of an Outbreak—How a 19 year-old Mexican man infected at least 56 others with tuberculosis and then eluded county health officials before being caught by police and quarantined."*[22]

An undetected illegal alien carrying TB may infect from 10 to 50 other Americans depending on public contact.

In February 2008, northwest Arkansas reported 100 cases of TB and nine cases of leprosy.

On November 6, 2003, at a local restaurant chain, Chi-Chi's in Beaver Valley, Pennsylvania,[23] unscreened employees 'served' up plates of infectious hepatitis 'A' to their patrons. Over 3,000 had to receive the painful gamma-globulin shots while two Americans died. Health officials reported, "Workers may have contaminated food by failure to follow basic hygiene in cleaning hands after using the bathroom." Owners failed to health screen employees at the restaurant chain.

Another distressing disease, leprosy, long feared from Biblical times, totaled 900 cases in the USA in the past 40 years. In the past seven years, according to a report from the NY Times in February, 2003, leprosy has infected over 7,000 people in the United States.[24] Illegal immigrants from India, Brazil, Mexico and the Caribbean brought leprosy into the U.S. Leprosy spreads by infected illegal aliens working in fast food, dish washing and hotels.

Chagas Disease arrives directly from Mexico and Latin America where it infects 18 million people. The T-cruzi protozoan destroys heart tissue and other organs. "One can contract it by eating uncooked food contaminated with infective feces of the Vinchuca Bug. It penetrates our border in the bodies of an average of 3,000 to 8,000 illegal aliens daily," said Carlos Bastien, author of the book: *Kiss of Death: Chagas.*[25]

Whether it's Dengue Fever, now in Florida, Hemorrhagic Fever[26] coming up from Texas border towns or E-coli intestinal parasites arriving with illegal migrants from Mexico daily, every American citizen faces these forms of bio-terrorism. The department of Homeland Security presents Americans with color coded 'alert' levels from Al Qaeda, but that doesn't protect us from a mounting invasion from an 'unarmed army' of disease carrying illegals that exceed 9/11 in mortality results.

You must demand your schools examine and process every child for head lice, TB[27] and leprosy on a regular basis.

SECTION V: INTERNATIONAL RAMIFICATIONS

CHAPTER 18: KILLING OUR OCEANS

"We're drowning in a sea of garbage with a three million ton waste dump twice the size of Texas floating 1,000 miles off the West Coast called the Great Pacific Garbage Patch."

Professor Paul G. Miller, Colorado State University

Humans consider the planet's oceans their private toilet. On average, they dump 8,000 pieces of plastic litter into the oceans and seas every 24 hours. The U.N. Environment Program calculated that 46,000 pieces of plastic litter float on every square mile of our oceans.[28]

Millions of Americans scoff at the idea that overpopulation causes any detrimental effects to our land, water and air. If they knew more, they might speak out and lead the *"Charge of Enlightenment."* They would push for change based on knowledge instead of emotions or past models. They might thrust aside anachronistic paradigms that no longer work in the 21st century by applying new concepts for present conditions.

Meanwhile, the August 1991 issue of Life Magazine, titled, *"Shark Alert!"* reported, "The age-old struggle between man and shark has become a killing frenzy. We slaughter 100 million sharks every year, driving many species to extinction."

Sharks prowled the seas for the last 400 million years. But now, the predator has become the prey. The Agger Trading Company in New York purchases shark fins where they are dried and resold. Hong Kong alone buys over seven million pounds each year for shark-fin soup.

Fast forward to the March 2006 issue of Mother Jones News, titled, *"Last Days of the Ocean."* Researcher Julia Whitty wrote, "One of the biggest culprits is long-lining, in which a single boat sets plastic line across 60 miles of ocean, each bearing ganglion lines that dangle at different depths, baited with 10,000 hooks designed to catch a variety of species. Each year, two billion long-line hooks are set worldwide primarily for tuna and swordfish—though long-liners inadvertently kill far more other species that take the bait, including 40,000 sea turtles, 300,000 sea birds and 100,000,000 sharks."

Not only that, fishing trawler captains cut loose thousands of miles of drift nets that get snagged on reefs. Those nets continue killing uncounted numbers of marine life by the millions for however long the plastic monofilament lasts: in a word—almost forever. Experts decry the practice of driftnets as 'clear cutting' under water where everything suffers destruction. Ecologists describe it as "raping the oceans with no moral or ethical responsibility."

Whitty continues, "Fishing fleets in the Gulf of Mexico have dropped the white tip shark population 99 percent since the 1950s, driving that species into virtual extinction. These sharks are thrown dead or dying back into the ocean; these unwanted species make up at least 25 percent of the global catch, as much as 88 billion pounds of life eliminated, for no reason, annually."

If you consider the figure of 100 million slaughtered sharks in 1991, annually, up to 2006, that's 15 years to slaughter 1.5 billion sharks for their fins for human soup. If you add in 300,000 seabirds annually, that's 4.5 million seabirds killed in those 15 years. That's over a half million sea turtles killed for nothing.

Whatever life remains in the Gulf of Mexico suffers from what is called a *'dead zone'* which is an area of ocean filled with human chemicals so toxic that few fish species can withstand it or reproduce. A deadly conveyor belt known as the Mississippi River delivers nitrogen-laced, chemically active fertilizers, herbicides and pesticides by the millions of gallons hourly, twenty four hours a day. Latest research shows a 10,000 square mile dead zone beginning at the mouth of Old Man River.

These examples represent humans' unspeakable power to devastate their ecosystems.

"Close to 50 hypoxic dead zones fester on the coasts of the continental United States," Whitty reported. "The situation is far worse in Europe, with 14 persistent dead zones that never go away, and almost 40 others occurring annually, the biggest and worst being the 27,000 square mile dead zone in the Baltic Sea, which equals the landmass of North Carolina."

The most sobering news, from my perspective, stems from human chemicals poisoning coral reefs. Because of climate change and chemicals injected into our oceans, the exhaustive study in 2004, *"Status of Coral Reefs of the World"* showed 20 percent of the world's reefs so badly damaged they are unlikely to recover and another 50 percent teeter on the edge of extinction…15 percent of the world's sea grass beds have disappeared in the past 10 years, depriving marine species of critical habitat."

"Likewise, kelp beds are dying at alarming rates; 75 percent are gone from Southern California alone—victims of the demise of sea otters that regulate populations of kelp-eating sea urchins."

Scientists at the National Oceanic and Atmospheric Administration estimated that oceans absorbed 118 billion metric tons of carbon dioxide since the onset of the Industrial Revolution. Today we add 25 tons daily. This exhaust stems from 84 million barrels of oil burned worldwide each day. That doesn't take into account the gas, wood and coal burned all over the planet—daily.

"This mitigation of carbon dioxide changes oceanic 'ph' levels," Whitty said. "Coral reefs plagued by so many stressors will almost certainly vanish."

A quick trip west of San Francisco during our nuclear expansion in 1945 showed the U.S. government dumped 400 barrels of radio-active waste 20 miles off shore. A recent PBS report showed all 400 barrels ruptured with their contents dissolved into the Pacific Ocean. Can you imagine what other nuclear countries have done with their radio-active waste? Is it any wonder tuna, salmon and other marine life test chemically positive and our children should not eat fish because their small bodies can't tolerate the doses deposited inside marine tissue?

As if our adult bodies fare any better! Read Dr. Sandra Steingraber's book, *Living Downstream,* for a sobering understanding of our chemically contaminated world and its cancerous consequences.

In Alaska, polar bears struggle and drown with vanishing ice pack because they must swim too far to gain the ice. Worse, toxic chemicals plague the ice bears. Researcher Marla Cone wrote, "Polar bear cubs already harbor more pollutants in their bodies than most other creatures on the planet. Mother polar bears store a lifetime of chemicals in their fat and bequeath them, via their milk, to their young. Several hundred of the industrialized world's most toxic chemicals like PCBs and organo-chlorine pesticides such as DDT have transformed the Arctic into a chemical repository. The chemicals magnify in animals each step up the food chain leaving polar bears, killer whales and other predators highly contaminated."

Today, in the Gulf of Mexico, leatherback turtles have declined 97 percent in the past two decades. They feed on jellyfish, but today, shrimpers can't draw their nets into the boats because millions of 25 pound jellyfish make it impossible to retrieve nets.

On top of this 'tip of the iceberg' report, you may appreciate that climate change causes horrendous storm activity in our warming oceans, so much so, hurricane scientists call for a new category greater than Katrina's category 5. They expect category 6!

If you noticed the fall hurricane season in 2008, Gustav, Ike and others lined up like bowling balls swirling toward the U.S. coast. With larger population centers, more people suffered disasters.

More tidbits to consider:

* Cruise ships produce 30,000 gallons of sewage and 19 tons of garbage daily which is dumped into the oceans in defiance of international laws.
* U.S. Congressman Richard Pombo (R-CA) has accepted over $23,000.00 in foreign junkets to help him weaken federal laws protecting fisheries and marine life.
* Gorton's Seafood has killed 2,700 whales against international laws in the guise of "scientific research."
* Toxic PCBs are regulated by the FDA. However, the FDA allows such high levels of cancer-causing PCBs in farmed salmon that, if they were present in wild salmon, the EPA would restrict consumption to one meal per month.
* Chemically caused cancers kill millions worldwide annually.

"Sewage—treated and untreated runs into the ocean from municipal and industrial plants, polluting water with bacterial waste, chemicals and metals," said Jane Kay, ecologist.

At what point in the thinking process does one's entrenched paradigm need excavation, lest a person or company or society bequeath on future generations a problem that proves both irreversible and unsolvable? Tell me there is no moral and/or ethical question as to this kind of killing spree by a cognitive species upon defenseless, non-cognitive fellow creatures!

Can Americans change course to a more viable and long-term sustainable future? If we continue our hyper-population growth, the answer is: no! As I noted in the first chapter of this book, you can't see a tsunami until it hits. As you read this information, it dawns on you that you couldn't see any of what you have just read, but it happens at an accelerating rate of speed—all caused by population overload.

How much more human expansion can we afford? At this point in the book, you know the answer!

CHAPTER 19: PLAGUE OF PLASTICS

"And Man created the plastic bag and the tin and aluminum can and the cellophane wrapper and the paper plate, and this was good because Man could then take his automobile and buy all his food in one place and He could save that which was good to eat in the refrigerator and throw away that which had no further use. And soon the earth was covered with plastic bags and aluminum cans and paper plates and disposable bottles and there was nowhere to sit down or walk, and Man shook his head and cried: "Look at this God-awful mess."

Art Buchwald, 1970

In my world travels from the Arctic to Antarctica, I found that humanity holds little sacred on this planet. I have sailed and used my Scuba gear across all the oceans and seas. I have rafted or canoed rivers from the Amazon to the Mississippi to the Yangtze. I have explored all the Great Lakes and many unknown lakes around the world. I have walked on the Hawaiian Islands to the Galapagos Islands to Ross Island at the bottom of the world. I bicycled along the North Sea in Norway and around Lake Titicaca in South America.

At every location on our globe that humans inhabit, humanity throws its trash in every conceivable form.

But by far the most dangerous—any way you cut it—plastics prove themselves humanity's worst invention. Ubiquitous, forever, deadly and ugly!

As a teenager, I enjoyed Scuba diving in pristine waters from Lake Huron, to the Hawaiian Islands, and from the Atlantic Ocean to the Caribbean. I saw magic at 40 feet below the surface on coral reefs! Incredible beauty!

Thirty years later, my dives carried me into the most disgusting sights on the planet. Plastic drift nets, cut away by fishing captains, killed innocent sea life--forever! For the past 40 years, humans have tossed their plastic containers, pop tops, disposable diapers, billions of bags and every kind and size of plastic trash into our lakes, rivers and oceans. Plastic destroys everything it touches.

As I canoed down the Mississippi River from its beginning at Lake Itasca, Minnesota, it started out as beautiful as a dream. Within five miles, I watched hundreds and then thousands of plastic containers float alongside me after having been pitched by other boaters. Plastic bags hung from trees and billowed in the water as they draped from branches along Old Man River. People drove cars over the river's edge and left couches and lawn chairs on sand bars. Clothes and junk got tossed along its 2,552 mile trip to New Orleans. It sickened me daily. I filled two large trash bags a day and I couldn't begin to get it all.

On my bicycle ride from Norway to Greece in 2005, we boarded a ferry from Brindisi, Italy to Petros, Greece. Along the way, we witnessed huge floating gobs of plastic trash collected in ugly swarms hundreds of yards long.

Plastic proves the worst human invention, besides chemicals, because plastic doesn't break down or biodegrade. About the only thing that destroys it is fire, but then, the pollution from the smoke proves fatal to the environment.

Alan Weisman, author of "Polymers are forever" published in the <u>May/June 2007</u> issue of Orion magazine: <u>http://www. orionmagazine.org/index.php/articles/article/270</u>

He wrote, "The true answer is we just don't know how much is out there."

Weisman wrote about Richard Thompson, "He knew the terrible tales of the sea otters choking on poly-ethylene rings from beer six-packs; of swans and gulls strangled by nylon nets and fishing lines; of a green sea turtle in Hawaii dead with a pocket comb, a foot of nylon rope, and a toy truck wheel lodged in its gut. His personal worst was fulmar bird carcasses washed ashore on North Sea beaches. Ninety-percent suffered plastic in their stomachs—an average of forty-four pieces per bird."

"There was no way of knowing if the plastic had killed them, although it was a safe bet that, in many, chunks of indigestible plastic had blocked their intestines. Thompson reasoned that if larger plastic pieces were breaking down into smaller particles, smaller organisms would likely be consuming them. When they get as small as powder, even zooplankton will swallow them."

"Can you believe it?" said Richard Thompson, one of the men researching how widespread plastic moved into water systems. "They're selling plastic meant to go right down the drain, into the sewers, into the rivers, right into the ocean. Bite-sized pieces of plastic to be swallowed by little sea creatures!"

If you are old enough to remember Dustin Hoffman in, "The Graduate", you may recall the older man telling Hoffman, "Plastic, my boy, that's the future!"

While WWII created research for plastics, after 1970, this unnatural substance changed everything and it became everything. Once it became a container, all hell broke loose. Every time a group of environmentalists tried to get a 10 cent deposit/return placed on it, corporations overpowered do-gooders with negative advertisements to defeat return laws.

Soon, the disposable diaper arrived! On my bicycle travels across America and the world, I've seen tens of thousands of soiled, plastic baby diapers thrown into every corner of the planet.

Weisman wrote, "What happens to plastic, however, can be seen most vividly in places where trash is never collected. Humans have continuously inhabited the Hopi Indian Reservation in northern Arizona since 1000 AD—longer than any other site in today's United States. The principal Hopi villages sit atop three mesas with 360-degree views of the surrounding desert. For centuries, the Hopis simply threw their garbage, consisting of food scraps and broken ceramic, over the sides of the mesas. Coyotes and vultures took care of the food wastes, and the pottery shards blended back into the ground they came from.

"That worked fine until the mid-twentieth century. Then, the garbage tossed over the side stopped going away. The Hopis were visibly surrounded by a rising pile of a new, nature-proof kind of trash. The only way it disappeared was by being blown across the desert. But it was still there, stuck to sage and mesquite branches, impaled on cactus spines."

On our oceans, "In 1975, the U.S. National Academy of Sciences had estimated that all oceangoing vessels together dumped 8 million pounds of plastic annually. More recent research showed the world's merchant fleet alone shamelessly tossing around 639,000 plastic containers every day.

"The real reason that the world's landfills weren't overflowing with plastic, he found, was because most of it ends up in an ocean-fill. After a few years of sampling the North Pacific gyre, Moore concluded that 80 percent of mid-ocean flotsam had originally been discarded on land."

Weisman wrote, "During his first thousand-mile crossing of the gyre, Moore calculated half a pound for every one hundred square meters of debris on the surface, and arrived at three million tons of plastic. His estimate was corroborated by U.S. Navy calculations. It was the first of many staggering figures he would encounter. And it only represented visible plastic: an indeterminate amount of larger fragments get fouled by enough algae and barnacles to sink. In 1998, Moore returned with a trawling device, such as Sir Alistair Hardy had employed to sample krill, and found, incredibly, more plastic by weight than plankton on the ocean's surface. In fact, it wasn't even close: six times as much."

As you read through this information, it gets uglier than you can imagine. "As for the little pellets known as nurdles, 5.5 quadrillion—about 250 billion pounds—were manufactured annually—perfect bite-size for little creatures that the bigger creatures eat, were being flushed seaward."

That half-century's total plastics production now surpasses 1 billion tons

Dear reader: as you can appreciate, it's what you can't see that produces incredible damage to our planet home. As I said, plastics prove the worst invention of humanity. They prove insidious, sinister, menacing and deadly to this planet's living creatures.

Any questions? Why would someone knowingly toss 639,000 plastic containers into our oceans daily? Why would Pete Coors, owner of Coors Brewing, Golden, Colorado tout himself as a Colorado environmentalist, yet spend millions of dollars to defeat our bottle/

return laws—not once but twice? Short answer: he personally makes $13 million a year, but that's too little! He requires more profit with total disregard for all the trash his cans, bottles and the plastic waste generate across the landscape. Just think of all corporation heads thinking and acting like Pete Coors. Sickening!

The next time you're at the grocery checkout, they may ask, "Paper or plastic?" You answer as you pull them out, "I've got my cotton bags, thank you." You may spearhead a 10 cent bottle deposit/return law in your state that duplicates Michigan's highly successful law.

It's a start. Nature thanks you!

For further insights into our ongoing onslaught of this planet, Alan Weisman's article is an excerpt from his book, *The World Without Us,* published by St. Martin's Press in July, 2007.

CHAPTER 20: BURGEONING CITIES OF POVERTY

"I don't think of myself as a poor deprived ghetto girl who made good. I think of myself as somebody who from an early age knew I was responsible for myself, and I had to make good."

Oprah Winfrey

Urban dwellers worldwide outstrip rural populations. Poverty, slums and pollution grow in the wake of population increases.

For example, each day, the United States grows by 8,200 people. Most of them settle into cities. California adds 1,700 people per day! One look at Los Angeles, Chicago, Houston, Atlanta and San Francisco illustrates hyper-population at its extreme.

How much of that total does legal and illegal immigration cause? Immigration fills two Pasadena Rose Bowls with 200,000 people every 30 days. We pour them into U.S. society and again refill the two Rose Bowls to pour them into America—month in and month out, year in and year out, decade in and decade out.

"Humanity will make the historic transition from a rural to an urban species sometime in the coming years, according to the latest UN population figures," John Vidal, author, *Burgeoning Cities Face Catastrophe.*

Vidal continued, "The shift will be led by Africa and Asia, which are expected to add 1.6 billion people to their cities over the next 25 years.

"The speed and scale of inevitable global urbanization is so great most countries will not be remotely prepared for the impact it will have," Thoraya Obaid, Executive Director of the UN Population Fund, said. "In human history we have never seen urban growth like this. It is unprecedented."

Ms. Obaid added: "In 2008, half of the world's population will be in urban areas. The shift from rural to urban changes a balance that has lasted for millennia. Within one generation, five billion people, or 60 percent of humanity, will live in cities. The urban population of Africa and Asia is set to double in this time."

She said that each week the numbers living in cities grew by nearly a million. This signifies world hyper-population growth on an epic scale. Makes you wonder what Barney Fife might say about that.

"Andy!" Barney said. "You ever wonder where common sense came from? You know how people figure somethin' out and make smart decisions?"

"Barn," Andy said. "When it comes to common sense, I ain't got humans figured out in this country let alone anywhere else on this here planet."

"Darn sure we need more common sense in these here times," Barney said, as he straightened his gun belt.

"Most cities in developing countries already have pressing concerns, including crime, lack of clean water and sanitation, and sprawling slums," Obaid said. "But these problems pale in comparison with those that could be raised by future growth. If we do not plan ahead it will be a catastrophe. The changes are too fast to allow planners simply to react. If governments wait, it will be too late to [gain] advantages for the coming growth."

Vidal said, "According to the State of the World Population Report, which Ms. Obaid launched yesterday in London, large-scale population growth will take place in the cities of Asia, Africa and Latin America. It suggests the largest transition to cities will occur in Asia, where the number of urbanites will almost double to 2.6 billion in 2030. Africa is expected to add 440 million to its cities in the same period, and Latin America and the Caribbean nearly 200 million. Rural populations are expected to decrease worldwide by 28 million people."

"But urbanization can be positive...no country in the industrial age has ever achieved significant economic growth without urbanization," said Obaid. "Cities concentrate poverty but they present poor people's best hope of escaping it. The potential benefits of urbanization, which include easier access to health centers and education, far outweigh the disadvantages."

"The report warns, however, that if unaddressed, the growth of urbanization will mean growth in slums and poverty, as well as a rise in attempted migration away from poor regions," Vidal reported.

"Today one billion people live in slums, 90 percent of whom are in developing countries," said Obaid. "The battle to cut extreme poverty will be waged in the slums. To win it, politicians need to be proactive and start working with the urban poor. The only way to defeat urban poverty is head on."

"Climate is expected to increasingly shape and be shaped by cities," Vidal said. "In a vicious circle, climate change will increase energy demand for air conditioning in cities, which will add to greenhouse gas emissions. It could also make some cities unlivable, adding to the 'heat island' effect, which can lift temperatures in urban areas by 2 to 6 degrees centigrade."

"Heat, pollution, smog and ground-level ozone [from cities] affect surrounding areas, reducing agricultural yields, increasing health risks and spawning tornadoes and thunderstorms. The impacts of climate change on urban water supplies are expected to be dramatic," the report says. "Cities like New Delhi, in the drier areas, will be particularly hard hit."

"What is taking place today, says the UN, is a second great wave of global urbanization," Vidal said. "The first, in Europe, from 1750-1950, boosted the numbers living in cities to about 420 million, but the second is expected to increase urbanization levels close to those found in Europe at 72 percent and the U.S. at 81 percent today.

"However, developing countries are at a great disadvantage when they start to urbanize," Vidal reported. "Mortality has fallen rapidly in the last 50 years, achieving in one or two decades what developed countries accomplished in two centuries. The speed and scale of urbanization today is far greater than in the past. In the first wave of urbanization, overseas migrations [to the US or Australia] relieved the pressures on European cities. Many migrants settled in new agricultural lands. Restrictions on international migration today make this almost impossible. They will also have to build faster than any rich country has ever done. It will require houses, power, water, sanitation and roads.

"The report also spells the end for growth of existing mega-cities. Only Dhaka in Bangladesh and Lagos in Nigeria, of the world's 20 mega-cities, are expected to grow more than three percent a year in the next decade. Most growth will be in smaller cities, of under 500,000 people. The good news is these cities are more flexible [in expansion]; the bad is they are under-served in housing, water, and waste disposal."

Ms. Obaid said, "It concerns everyone, not just developing countries. If we plan ahead we will create conditions for a stable world. If we do not, and do not find education, jobs, and houses for people in cities, then these populations will become destructive to themselves and others."

As you can plainly see in this report, accelerating populations create irreversible crises with unsolvable problems. For Ms. Obaid to even conjecture that something good will come out of such massive urban expansion shows a lack of understanding of climate change, species extinction, air pollution, acid rain and the mounting aggregate of consequences tied to hyper-population expansion. She fails to understand exponential growth!

Yet, her brand of thinking remains pervasive in the face of cascading consequences on a global scale. When you read any economic report, the one factor most applauded remains that 'obese' word—growth.

By applying worldwide illustrations, you can see what will happen to America with an added 100 million people. It becomes more daunting when you realize 70 million of the next 100 million people added to the USA comprise third world immigrants escaping their own dilemmas worldwide—only to recreate those population problems in the United States.

Los Angeles cannot survive another 10 to 15 million added people, but that's where they're headed. If you remember, the arrogant Titanic steamed along in iceberg filled waters in 1912. Nothing happened and all proved well until it hit the iceberg. Once it hit, all hell broke loose. Everyone became victims and/or survivors. Once 100 million people manifest in America, we too, will become like Dhaka, Bangladesh or Beijing, China or Bombay, India. We've already seen what they've done to themselves. We do not have to do it to our children.

When will leaders inject concepts like: "International Sustainable Population Policy"; "International Carrying Capacity Policy"; "International Water Usage Policy"; "International Environmental Impact Policy" for all countries?

Once they determine *carrying capacity* for their geography, citizens may enjoy a sustainable existence. From that equation, "International Family Planning" enters the picture.

CHAPTER 21: ENVIRONMENTAL REFUGEES

"Anyone who believes exponential growth can go on forever in a finite world is either a madman or an economist."

Kenneth Boulding

With continued U.S. population growth rates, California tops an added 18-20 million by 2035. Texas grows by 12 million by 2025. Every state shares negative ramifications of the next added 100 million people.

Some experts at Vanderbilt University tell us that U.S. population stands at 327 million people, and the growth-rate accelerates even faster, so we will reach 400 million sooner—by 2025-2030.

As you can see, California, already at 37.5 million, absorbs the lion's share of this population "Human Katrina." At each stage along the way, as those 20 million additional people impact that state, can you imagine what will happen? Will people fight one another for shrinking natural resources?

How about water? How about clean air— California smog? How about farmland? How about traffic? How about housing? How about schools? How about standard of living? How about quality of life? How about greenhouse gases? How about grid-lock?

Ever hear the sound of a toilet flushing? Ever see one backed up? Ever see and smell the results of both? How many times can you expand a sewage-treatment plant? Where do you deposit trash? There you have it! California population-density in twenty odd years!

Environmental Refugees

What can we expect from adding 100 million? Call them 'environmental refugees', for lack of a better term. You can bet your bottom dollar they're coming to America. If you doubt me, just travel to the Mexico border, and watch the "pulgas" that come across illegally every night. The Border Patrol only catches 20 percent.

Human beings become environmental refugees when they exceed the carrying capacity of that portion of the planet that supports them.

Religion and population-growth impacts

For better or worse, the Catholic, Hindu, Buddha and Islam churches encourage unlimited birth rates. With more numbers come more power, and more money for the churches.

If you look at the results on the world stage, environmental refugees stream into Europe from Africa and the Middle East. If you look at Central and South America along with Mexico, those same refugees stream into the U.S. at an accelerating rate of speed.

Can realities like starvation and deaths from overpopulation change religious doctrine based on books written 2,000 years ago? Not likely! Even in the face of severe environmental consequences, do not look for the Pope to change his stand. Neither will any Muslim leader stand up against 1,400 years of doctrine! Ancient religions lock followers into the past as well as the familiar. Humans tend toward the security of their habits.

The most frightening aspect of the next added 100 million Americans stems from the fact that they won't be Americans. Do they speak English? Which national flag hangs in their home in the U.S.? Where does their primary allegiance rest?

As Mark Krikorian in his book, *The New Case Against Immigration: Both Legal and Illegal,* said, "Today's immigrants have one foot in the United States and their ear, via cell phones, back in their old country. They never fully integrate into American society. They constitute 'transnationals' with no single home country or loyalty."

What about overloaded populations?

Mexico grew from 50 million to 104 million in the last century. At the minimum, Mexico expects 153 million people by 2050 if not sooner. If they can't feed, clothe and house countless millions today, what can we expect when they add 50 million?

Why is the U.S. the lifeboat to the world's unfortunates?

No less than 30 to 40 million destitute Mexicans—if they can't feed, water and clothe themselves—will move out of their country in the 21st century into the United States. Over 12 to 15 million Mexicans have already moved into the U.S. As their conditions worsen in their home country, millions more will migrate. Major portions of Central America will migrate with the Mexicans, leaving similar situations.

"The vision of a world teeming with environmental refugees is daunting, even for wealthy countries such as the United States," said John Cairns, University Distinguished Professor of Environmental Biology at Virginia Polytechnic Institute and State University, Blacksburg, Virginia.

"When countries capable of absorbing refugees are at or beyond their carrying capacity, every individual on the planet becomes a potential refugee with no place to go," Cairns said. "The world's human population is increasing at 77 million net gain per year, but natural support-systems are decreasing."

Ecological footprint growing dangerously

"It's abundantly clear that many countries have exceeded their carrying capacity, and can only survive by exporting surplus populations," Cairns said.

Additionally, the United States, with five percent of the world's population, creates 25 percent of greenhouse gases. Will we take responsibility for environmental refugees flooded out of their countries or islands by rising oceans from climate change? That number could reach into the hundreds of millions.

Climate change--ocean levels rising-- resource-competition wars

Simultaneously, what about our own refugees if climate change floods Florida? Some climate models show one-third of Florida under water within this century. That could mean as many as 10 million refugees.

"It is becoming increasingly probable there will be teeming millions of environmental refugees—it will mean that we were terribly wrong about the carrying capacity of the world for our species," Cairns said. "Disease, starvation and resource-competition wars will doubtless occur—but the root cause will be a social disequilibrium resulting from overpopulation and bad long-term management of natural capital."

Do you understand that when a third world person reaches America, their 'ecological footprint' changes from near zero to 12.6 as soon as they take a job, buy a car, rent a house, buy goods and produce waste? This occurs into the millions with legal and illegal immigration annually.

Environmental refugees that survive will converge on America and other successful Western societies at all costs.

We can awaken—stop the onslaught—but, will we?

We can change the environmental refugee equation:

1. When world leaders accept that infinite population growth on a finite planet is not possible.
2. When humans realize they are dependent upon this planet's ecological support systems. They cannot continue to damage the systems without suffering severe consequences.
3. When leaders understand that achieving sustainable use of the planet will not be possible if human population and/or per capita consumption continues increasing.
4. When humans can't or won't come to balance, nature's means of famine, disease, wars and chaos will do the job brutally. (Cairn)
5. When birth control creates human stabilization on a finite planet.

The most astounding aspect of this data—that ties my mind in knots—stems from the fact that the average citizen doesn't seem to have a clue about what's coming—and not one national leader will address it.

Why? Money! Follow the money. Every time any environmental shock wave surfaces—like climate change, fisheries collapsing, species extinctions and oil depletion—our leaders think of a dozen ways to obfuscate, deny or ignore our accelerating calamity. Meanwhile, citizens go about their business completely clueless.

Are we willing to change?

Can we stem this impending environmental refugee flood? We better! We are not the world's lifeboat.

First world countries may provide economic assistance to overloaded countries. Most impactful, advanced birth control stops the cycle of poverty and overpopulation. Citizens of the world must be helped in their own countries.

While presenting my program, "*The Coming Population Crisis in America: and what you can do about it*", a college student asked, "How can you force other countries to practice birth control if it's not a part of their culture?"

"Great question," I answered. "We can't force them to use birth control. We can only offer it. If they choose not to use it, then, as first world countries close their borders to hyper-population growth nations, it will force them to deal with their own populations one way or the other. Either they choose action or nature takes action. That may sound harsh, however, it will be just as harsh for overloaded first world countries when the same population dilemmas ultimately manifest within their borders."

As in the past, until another "Hurricane Katrina" knocks American society upside the head with a whopping sledge hammer—cultural and social changes of the magnitude illustrated in this book— won't be heeded. While Americans ignore nature—nature will not ignore America.

The United States tiptoes along the edge of a dangerous cliff. Americans skipped along that cliff for the past 50 years. At least two billion of the earth's humans live on less than $3 each per day. Very few Americans comprehend that kind of existence. I've seen it and it's ghastly, depressing and growing by, well, you know!

We see environmental refugees stream into America from Mexico, Central America, India, China, Africa and a hundred other countries. They've added 106 million people to our country in the past 43 years. Are you ready for the next 100 million in 30 years?

The rate of growth increases exponentially.

What preparations have you made? Are you coaching your kids as to what they face? Have you Googled for the nearest walled city? Have you spent your entire 401K toward property with abundant water in the northeast corner of Montana? If you hail from Florida, how about adding 10 pounds for the nine-month winter? Have you cut a good store of wood? Can you drive horses behind a one furrow plow? Have you completed your class on how to can corn, green beans and blackberry jam? Did you order wall paper and insulation for your outhouse?

For more ideas, visit www.transitionus.ning.com

CHAPTER 22: COMING MEGA-TRAUMAS

"Creation destroys as it goes, throws down one tree for the rise of another, but ideal mankind would abolish death, multiply itself millions upon millions, rear up city upon city, save every parasite alive, until the accumulation of mere existence is swollen to a horror."

D.H. Lawrence

If the first sections of this book failed to alarm you concerning your children's future, I scratch my head in dismay.

I slap my jaw in distress realizing our nation's president and Congress carry on their inane activities without a clue or action toward a viable future. A few visionaries like Tom Tancredo, Dr. Albert Bartlett, former Colorado Governor Richard Lamm, Dr. John Tanton, John Rohe, Roy Beck, Barbara Coe, Terry Anderson, Dr. Diana Hull, Lester Brown, Kathleene Parker, Barbara Jordan, William Gheen, Jason Mrochek, Fred Elbel, Haydee Pavia, Priscilla Espinoza, Lupe Moreno, Sharma Hammond, Rosemary Jenks, Dan Stein, Beth Thomas, and a growing army of Americans do understand, however, the power elites drag their feet. Most Americans follow the president and Congress into our growing population mega-trauma.

What mentality defines unending growth as progress?

Short answer: economists, politicians and fools!

To support that, Covert Bailey said, "The brain is entirely made of fat. Without the brain, you might look good, but all you could do is run for public office!"

Author Joseph Tainter, *The Collapse of Complex Societies*, demonstrates that collapse is a frequent, if not universal, fate of multi-faceted societies. Collapse relates to declining returns on efforts to support growing levels of complexity with diminishing supplies of energy extracted from the planet.

Collapse: How Societies Choose to Fail or Succeed by Jared Diamond, illustrates how civilizations in the past failed because of resource constraints.

By 2050, at our accelerating rate of resource use and abuse, Jared Diamond's son or grandson might write a book titled: *How the United States Flushed Itself Down the Toilet by Its Own Hand.*

Additionally, author Richard Heiberg, *Peak Everything: Waking Up To A Century of Declines,* lists five axioms on sustainable or unsustainable civilizations:

1. Joseph Tainter's Axiom: any society that continues to use critical resources unsustainably will collapse.
2. Dr. Al Bartlett's Axiom: population growth and/or growth in the rates of consumption of resources cannot be sustained.
3. To be sustainable, the use of renewable resources must proceed at a rate that is less than, or equal to the rate of natural replenishment.
4. To be sustainable, the use of non-renewable resources must proceed at a rate that is declining, and the rate of decline must be greater than or equal to the rate of depletion.
5. Sustainability requires that substances introduced into the environment from human activities be minimized and rendered harmless to biosphere functions.

Do any U.S. leaders present anything resembling solutions to these factors to maintain a sustainable American future? The obvious answer: no! We burn 20 million barrels of oil in the USA with an additional 64 million barrels worldwide per day. We continue consumption rates beyond our finite resource base. We inject 72,000 chemicals into the air, water and land daily along with endless plastics tossed into our oceans.

Where do our actions lead?

Governor Lamm, a man I know personally and respect highly, wrote, *Megatraumas: America at the Year 2000.* He projected ramifications of our growing consequences from population growth. Even if he was off by a few years, he hit the bull's eye faster than

Deputy Barney Fife could pull out his pistol to stop the bad guys in Mayberry!

"The U.S. economy will be debt-ridden, with structural unemployment nearing 20 percent," Lamm said. "The U.S. will have the lowest percentage of capital investment and lowest growth in productivity and savings of any major industrialized country. The middle class will be wiped out by these inter-related economic predicaments."

A quick look at the Fannie Mae, Freddie Mac, AIG bailouts in the later part of 2008 proved small potatoes compared to the $700 billion bailout from the mortgage loan collapse in September of that year.

In 2009, we suffer $9.4 trillion national debt, $789 billion devastating war debt, 14 million unemployed Americans, 28 million Americans on food stamps,[29] 3 million jobs outsourced, one million insourced and millions more offshored. We borrow $2 billion daily from foreign investors to float our economy. Consumers suffer a gargantuan $2 trillion debt. Our manufacturing base no longer exists and illegal workers from Mexico create the second largest underground economy in the world. According to Bear-Stearns Report of 2005, $400 billion in IRS income taxes go uncollected while American job wage losses exceed $200 billion annually.

"The U.S. has the most expensive and inefficient health-care system in the world," Lamm said.

Because of massive immigration numbers, 86 hospitals and ER wards in California bankrupted in the past five years. Over 40 million Americans cannot afford health insurance in 2009.

"Immigration will be out of control," Lamm said. "At a time when the U.S. economy was not creating enough new jobs for our own citizens, the federal government allowed millions of legal and illegal immigrants to enter. The Southwest will become another Hispanic Quebec.

"Taking a cue from Castro's success in 1980 when he released thousands of criminals into the hands of Florida, other countries sent terrorists and criminals across the U.S. borders."

We suffered 9/11. Our prisons hold 400,000 illegal alien convicts costing $2 billion annually. American citizens suffer death daily at the hands of illegal alien drunks, killers, rapists, MS-13 gangs and children molesters.

"America's educational system will move into a new Dark Age," Lamm said.

In Lamm's state of Colorado, Denver Public Schools suffered a 67 percent drop out/flunk out rate in June 2005. One of five teachers quits or transfers out every nine month cycle from the futility of chaotic classrooms. In California, education turned into a fight for survival with 100 competing languages, gangs, drugs and racial conflicts. Educators 'warehouse' immigrant children until they dispatch them into our society at age 18—functionally illiterate.

In June 2008, NBC's Brian Williams reported a 76 percent dropout rate in Detroit, Michigan schools. Dozens of big city school systems reported similar figures for their failing schools.

"About 1.2 million teens hit the job market every year without a high school diploma," Williams said.

Washington Post journalists reported that one third of the Washington, DC's local population suffered functional illiteracy while one fifth of mainstream America can't read, write or work simple math.

"America will become a dangerous, crime-infested country," Lamm said.

One look at 2.3 million prisoners in our federal, state and local jails tells you something amiss in America. Over 15,000 MS-13 gang members distribute $100 billion in drugs to our schools and cities nationwide.

"Our groundwater will become badly contaminated," Lamm said.

Over 40 percent of lakes and rivers in America currently remain unsafe for swimming or drinking. The Mississippi River spews toxic water into a 10,000 mile dead zone[30] in the Gulf of Mexico.

"The U.S. carries an international welfare load to match its domestic one," Lamm said. "An effort to increase exports to 'never-to-be-developed' countries has led to a dangerous overexploitation of American farmland."

The U.S. gives over $6 billion annually to countries like Israel and Egypt in outright money that goes into the pockets of the rich. It pays horrendous welfare to millions in the USA—unable, unwilling or uneducated enough to work.

As shown by Dr. Tanton's research in *"Crossing our Agricultural Rubicon"*, as a nation, we no longer enjoy unlimited food production as our farmland diminishes with population onslaught.

As you read this book, you discover numerous ramifications concerning population growth. It's like gazing across a beautiful lake that appears pristine on the surface until you dive under water to see rusting cans of nuclear waste, leaking cans of pesticides, chemical piles, wrecked cars, dead fish on the bottom and a host of other ugly realities.

Lamm in the 1980s saw America as a "nation in liquidation"

Lamm felt we might hold out hope with a brilliant new president who leads the country back to a sustainable future. In his book, fictional President Morgenstern ("Morning Star") managed to balance the budget, paid off the debt and created means-testing for welfare applicants. National health showed improvement by prevention methods; pension systems overhauled; government regained borders; infrastructure rebuilt; and economic aid limited to countries that helped themselves.

In light of what's happening today, I am curious as to Governor Lamm's projections in 2009.

The late Alexander Solzhenitsyn warned, "Seen from the outside, the massive upheaval in American society approaches a limit beyond which it will become 'meta-stable' and must collapse."

Even that famous Russian writer who lived here for a short time could not imagine the environmental aspect; nonetheless, he witnessed the social and historical ramifications.

If you look at what our president and Congress continue in 2009, I add 'national suicide' via apathy as our ultimate destination. Most

Americans resemble sheep following the one before them down a chute into the killing zone. No one takes action or tries to escape until they see the sledge hammer dropping toward their head. By then, it's too late.

Every sign points to environmental breakdown, population pileup and degraded standard of living. Like Hurricane Katrina, you can count on the consequences with continued American mass apathy as well as incompetence and corruption of elected leaders.

CHAPTER 23: EXAMPLES OF
OUR FUTURE ABOUND

"We must alert and organize the world's people to pressure world leaders to take specific steps to solve the two root causes of our environmental crises - exploding population growth and wasteful consumption of irreplaceable resources. Over-consumption and overpopulation underlie every environmental problem we face today."

Jacques-Yves Cousteau

In 1950, Bangladesh grew toward 100 million people. Today, it overflows with 144 million people—expecting to double its population by mid century. That might not be so bad, except for one thing: Bangladesh endures those numbers in a landmass less than the size of Iowa!

Dr. Alan Kuper, recently deceased, a man I admired and respected, illustrated our accelerating dilemma with metaphors that make sense. He reported from his web site using the research of John Wenzel, director the Ohio State University Museum of Biological Diversity. Wenzel said, "In the days of sailing ships, sailors used to leave goats on islands as they passed to ensure fresh meat on return trips. It worked too well. The animals bred faster than the sailors could eat them, and from the Channel Islands off California to the Seychelles in the Indian Ocean, goats ate all the vegetation and starved. The goats also screwed up the environment so that native species couldn't survive, either. For example, the goats stripped away the plants' low-growing leaves so that tortoises couldn't find enough to eat."

"The Inter-governmental Panel on Climate Change released its report blaming humans for increased temperatures, melting glaciers and rising seas," Kuper's website reported. "Too many people burn too many fossil fuels at 84 million barrels daily."

That doesn't account for millions of tons of coal, natural gas and wood burned every day by 6.7 billion humans.

"With global warming, we've been able to create this problem in the first place because we've had virtually free energy in the form of fossil fuels," said Ohio State University ecologist Tom Waite.

Kuper reported, "Climate change, Waite and others say, is a sign that we are exceeding the number of people Earth can sustain. Every year, at least 77 million humans are born in excess of those who die. That's 1 billion people every 11 years. Some, however, argue that we are adept at adapting, and point to increased agricultural production and medical advances that fend off disease."

Right now, Earth's carrying capacity is thought to be somewhere in the range of 1.5 billion to 2.0 billion people with a Western standard of living. We sport 6.72 billion today and grow by 240,000 every 24 hours as we add 77 million annually. Visit www.populationmedia. org[31] to see the population meter.

You might be curious about the above paragraph. In other words, we're already out on a hazardous limb as we have overshot the world's carrying capacity in a short 100 years. Michael Brownlee of www. transitionus.ning.com , said, "We face a perfect storm with 'Peak Oil', 'Climate Change' and 'Economic Instability'. China and India want what we have but they won't be able to obtain it. We want to keep what we have but it won't be possible with oil's decline."

At some point, the human race must lower its population on this planet by birth control and family planning—gracefully—or nature will bring its own brutal methods to the table with a greater vengeance than the 18 million who starve to death currently on this planet each year.

That might lead you to ask the carrying capacity for humans in the United States. If we would like to continue our current standard of living, Cornell Professor David Pimentel projected 150 million. If we lowered our energy usage, resource usage and cut in half our 'ecological footprint' to 6.3 acres—obviously we could support more people. But then, what's the point? Why push the envelope with so much damage and consequences guaranteed with larger human numbers?

"In biology, the carrying capacity usually refers to the number of animals a given area can support with adequate food, shelter and territory or the space to reproduce," Kuper reported. "Duke

University ecologist Stuart Pimm said half of the world's population has little access to medicine, electricity, safe water and reliable food supplies."

"When you get to the nitty-gritty of the term, some animals are more equal than others. Some countries are a lot more equal than others," Pimm said. "You might have 50 billion, but the quality of life might not be terribly pleasing. Rabbits are the same way. The key word is support."

"The United States possesses resources to sustain less than half of its current population of 306 million, according to ecologist Paul Ehrlich, who first called attention to potential population problems in 1968 with his book, *The Population Bomb*, Kuper reported. "Waite and other ecologists increasingly think of the idea of carrying capacity in terms of an "ecological footprint," the amount of land on Earth that it takes to support a group of people."

Americans, who make up 4.8 percent of the world's population, use 25 percent of its resources and cast a large footprint.

"Ohio's footprint is like 11 times the state of Ohio," Waite said. "The average American's footprint is about 12.6 acres. By far, the largest component is energy. In contrast, the average citizen of India has a footprint one-sixteenth that size."

Kuper reported, "If all 6.7 billion people were to share the world's resources equally, Americans would have to reduce consumption by 80 percent for each of us to have a footprint of about 4.4 acres."

"Waite rides a bicycle most days. However, Waite's footprint grows when he arrives on campus or has to rent a car to attend a meeting out of town," Kuper reported.

"The moment I show up at work, I suddenly violate the fair Earth share and I become unsustainable," Waite said. "Carrying capacity and footprint are tied to the global economy, which has quadrupled since the world's population doubled."

"That leads to a fear that slowing population growth might not ultimately curb greenhouse gas production if more people achieve Western lifestyles," said Barry Rabe, a University of Michigan political scientist who studies environmental issues. "India and China are developing rapidly and have already affected climate change. China is opening an average of one coal-fired power plant a week to meet electricity demand. The power plants emit carbon dioxide, a greenhouse gas."

"Everyone in China wants two things: their own apartment and their own car," Pimm said. "That change is going to have a massive effect on the planet."

Lester Brown, President of the Earth Policy Institute, said the sustainability question is a loaded one, "People ask me how many people the Earth can sustain. I usually respond that it depends on whether you want to live like an Indian or an American."

"For example, farmers worldwide grow about two billion tons of grain every year. Each American consumes an average of 1,760 pounds annually, mainly because of the grains used to feed farm animals. If everyone on the planet consumed that much grain," Brown said. "Earth would support about 2.5 billion people. But in India, people consume about 440 pounds each. If everyone else in the world did likewise, the world's grain would support about 10 billion people."

Population, water and food lock horns. Growing one ton of grain requires 1,000 tons of water. Water shortages already occur in Africa, Asia and the Middle East. For a sobering reality check, water shortages already occur in the United States of America! As water flows from agriculture to support growing urban populations, more grain must be imported.

"Alternative energy, touted as a possible solution to burning fossil fuels that generate greenhouse gases, also adds a factor to the food equation," Kuper reported. "Soybeans are increasingly in demand for biodiesel. And ethanol production now vies with food for corn. Brown estimates that half of the U.S. corn crop will go to ethanol."

"Seventy percent of all corn imports in the world come from the U.S., so what happens to U.S. corn crops affects a lot of countries," Brown said.

Waite said this competition for energy and food will change the landscape, "If we were to replace our reliance on fossil fuels and instead grow fuel plants, which would require setting aside lots of land to produce ethanol. We don't have enough land worldwide to meet those demands. Demand for food, fuel and materials already consumes more trees and crops than are grown worldwide."

115

Waite compares the issue to a bank account, "Humans are already drawing on capital rather than interest, and once that is exhausted, they will find Mother Nature reluctant to make a loan."

As you surmise on the next added 100 million Americans, we craft a mighty deep chasm for our children. Check out www.uscongress-enviroscore.org for a greater understanding of what we face.

CHAPTER 24: OVERLOADED NATIONS

"In the decades ahead we will be going through hell. That is an awful thing to contemplate, but the only alternative to accepting the fact is to live in denial until the reality is inescapable and our room for maneuvering is even more restricted than it has already become. What we must do now is lay the groundwork for collective survival."

Richard Heinberg, Peak Everything

If you don't think we're in trouble—I seize this opportunity to bring your attention to other specialists who understand our situation.

They've seen what I've seen. They know what I know. They understand: when you 'see' it, you comprehend it. Until you have seen it, you can deny or ignore it.

Refresher: the 'official' U.S. Census Bureau told us that we 'celebrated' three hundred million Americans in October 2006. Meanwhile, Vanderbilt University reported the actual number exceeded 327 million.

In the U.S., we overload demographic saturation levels more than average citizens realize. Would our government intentionally understate demographic numbers?

Nothing about the next added 100 million Americans can be ignored without accelerating our decline. When the "new folks" manifest upon our shores, a magic wand cannot make them vanish.

Lawrence Smith, past president of the Population Institute, (www.populationinstitute.org) wrote for the Providence Journal on the overloaded populations developing around the world.

Speaking at a symposium in the National Press Club, July 2007, Smith said, "The eminent environmentalist Lester Brown said he was pondering a question I don't believe he, or anyone else, really wants answered: "How many failed nation-states would it take to make a failed world?"

The World Bank, which prefers to call them 'fragile nation-states', recently identified 26 countries that pose some of the world's "toughest developmental challenges," noting that all face similar hurdles:

* Weak security
* Fractured societal relations
* Corruption
* Breakdown in the rule of law
* Lack of mechanisms for generating legitimate power and authority

These countries already experience massive human die-offs from famine and disease that we discussed earlier. Refresher: eight million adults as well as 10 million children die every year from starvation globally.

Smith continued, "Poor governance and extended internal conflicts are common among these low-income countries under stress, a new World Bank report observes, but past international engagement has failed to yield significant improvements. The report emphasizes that to avoid adverse spillover effects—such as conflict, terrorism and epidemic disease—the international community and the World Bank need to find more effective ways to assist these fragile states."

Fragile nation-states freefall -- horrible consequences

Nothing can be done once these fragile states freefall into starvation, wars, terrorism and genocide. Witness Darfur and Sudan today.

Do you think Americans care? Short answer: nope!

However, given enough time, the same problems impacting those societies migrate to our society.

"Former U.S. Secretary of State Condoleezza Rice warned that failing countries presented 'unparalleled danger to U.S. security' and that they 'serve as global pathways that facilitate the spread of pandemics, the movement of criminals and terrorists, and the proliferation of the world's most dangerous weapons,'" Smith said. "Though the World Bank report points out that the countries it cites are home to nearly 500 million people, roughly half of whom earn less than $1 per day, it does not single out spiraling human growth as a factor in their plight."

Today, two billion people stand at-risk on planet Earth. Third world nation inhabitants dream of migrating to a first world nation. This thinking brews unrest and contempt.

Look around this world as to demographic overload

You can see, read or hear about it on any public service channel. You'll see religious leaders and programs to 'feed the children'—but when they feed the children; those they feed grow up to birth even more children—kids that die in greater numbers.

If they used common sense, they would stop facilitating burgeoning populations and instead, balance their compassion with the provision of birth control and education along with food.

If they don't provide birth control, those organizations generate greater deaths in the future. Again, you cannot fool Mother Nature.

"The total world fertility rate is 2.7 children per woman; for the industrialized world, it is 1.6 children per woman," Smith said.

* The omission of rapid population growth from the World Bank's report is a serious fault -- considering that women average six or more children in eight of the countries it labels as fragile: Afghanistan, Angola, Burundi, the Democratic Republic of Congo, East Timor, Guinea-Bissau, Liberia and Somalia.
* The birth rate is five or more children in another eight countries: Congo, Ivory Coast, Eritrea, Guinea, Nigeria, the Palestinian territories, Sudan and Togo.
* And it is four or more in six fragile states: Central African Republic, Comoros, Solomon Islands, Vanuatu, Haiti and Laos.

Smith said, "Such high total fertility rates lead to disproportionately large youth populations—an indicator, in impoverished countries where educational and employment opportunities are few or nonexistent—make fertile ground for radical and terrorist group recruitment."

Please soak this in

World population grows to a high of nearly ten billion or more within the next four decades. This number haunts future generations— as they contemplate what we could have done to stop the population juggernaut. I suspect the eighteen million human deaths annually from starvation will be five to ten times greater.

"Two-thirds of the failing countries are projected to have population increases of 118 percent or more by mid-century, a nightmare scenario considering that they are already mired in the quicksand of poverty and deprivation—and resistant to rescue efforts by the global community," Smith said.

Africa, at 767 million people today, expects to double its population in this century. Mexico grows from 106 million to 153 million in 40 years.

Lawrence Smith talks about a global illusion – belief in magic, smoke, and mirrors

I would go one step further and say the United States suffers from a national illusion and/or national denial of reality—a typical view held by addicts. The U.S. hangs itself on the gallows of unending population growth.

As you look around, again, no leader in the US, no governor, no senator or anyone in power speaks to this national crisis. No world leaders touch it!

The fact is—it's coming as surely as the dawn

"The failure of the World Bank to mention, much less discuss, rapid population growth as a root cause of the conditions engulfing virtually all fragile countries reflects a mindset in international circles," Smith said. "Population stabilization fades as a development priority, perhaps because fertility declines in much of the industrialized world have created the false impression that rapid human growth is no longer a critical global concern."

Smith continued, "This illusion neglects to take into account the soaring human numbers in the poorest of the poor countries, where the provision of food, shelter, health care, education and employment is all too often problematic at best."

Ironically, safe and effective family planning methods exist, but the political courage to offer them falls far short of the need. QS Permanent Female Contraception (www.quinacrine.com) provides a safe, inexpensive permanent birth control method.

Addiction to population growth blinds us

Smith finished with, "Meanwhile, the World Bank's list of fragile states has expanded from 17 to 26—a 53 percent increase in only the last three years."

No one pretends that substantially curtailing human growth will be the salvation of these troubled nations, but to preclude the population issue that must be factored into the equation, remains a glaring oversight.

CHAPTER 25: CLIMATE CHANGE STARTS

"I don't mean to imply that we are in imminent danger of being wiped off the face of the earth - at least, not on account of global warming. But climate change does confront us with profound new realities. And unless we do a better job of adjusting to these new realities, we will pay a heavy price. There will be a toll on our environment and on our economy, and the toll will rise higher with each new generation."

Eileen Claussen, speech, July 17, 2002

We could sit back in the old Lazy Boy lounger; press the remote, drink a beer and watch our population increase to unsustainable levels. We may think 30 years extends so far into the future that it's not important to take action today. What the heck! One hundred million more Americans won't be any different than the last 100 million we added since 1965. Since we now house 306 million, what's the problem with another 100 million?

Short, sobering and gut wrenching answer: plenty!

You may find it interesting to note in 1906, India reached 300 million people. Today, at four times our U.S. population, demographic experts say India, at 1.1 billion, will out-grow China on India's way to 1.5 to as high as 1.6 billion people by mid century. Both China and India have become SO accustomed to degraded living and human misery, they endure it as normal.

As the human race speeds toward its mid-century destiny, I invite you to visit numerous population websites showing the number of people added to the planet every minute. If you're drunk, it will sober you up. If you're having coffee, it will sour your cream. If you add sugar, it will taste bitter. If you raise children, you will wonder what awaits them.

But long before we make it to mid-century; and long before the United States reaches that next 100 million added citizens; and long before India and China add another 500 million each—all of us must contend with another threat to our existence in the 21st century: climate change.

Before we get started, let's understand climate change versus global warming. Let us appreciate that climate change stems from a disruption of or an imbalance in our planet's ecosystem. Climate change may be a better term than "global warming." Additionally, it may be a natural cyclical phenomenon. The fact remains that climate change manifests across the planet in the early part of the 21st century.

Yes, you'll hear naysayers and you'll hear excuses, but I spent a season 'on the ice' in Antarctica where I saw climate change in action. I watched the best scientists in the world report their findings. Folks, we face daunting challenges! If you remain convinced that humans do not cause climate change, that's fine, but it's manifesting and it's accelerating faster and faster—demanding mitigation or preparation for the consequences.

However, some scientists negate the findings of other scientists. Professor of Atmospheric Science at Colorado State University said, "I have closely followed the carbon dioxide arguments. From what I have learned as to how the atmosphere ticks over forty years of study, I have been unable to convince myself that doubling of human-generated greenhouse gasses can lead to anything but quite small and insignificant amounts of global warming." He could be right, but if he's wrong, why should we err on the side of continued population expansion?

From simple common sense, we cannot expect to burn 84 million barrels of oil every 24 hours; we cannot burn billions of tons of coal for electricity annually; we cannot burn millions of metric tons of natural gas daily; we cannot burn millions of board feet of wood every day worldwide—and consider the planet's delicate ecosystems immune. If your reasoning follows that course, this chapter and the following chapter offer greater understanding from the experts.

As it stands, the U.S. remains the only country without its signature on the Kyoto Treaty regarding mitigation of climate change by reducing carbon emissions.

It matters little whether you like former Vice President Al Gore or not, but his movie, *"An Inconvenient Truth"* awakens even the idiots among us. Those who counter his movie, from my first hand knowledge in Antarctica and around the globe—illustrate and promote their own form of intellectual illogicality.

Today, we resist the facts; we deny their proven reality; we recoil at what faces us. Most U.S. citizens cower in a corner concerning the reality of climate change. We resist partnering with other nations in finding solutions. We continue accelerating our population which adds momentum to our impending Katrina-like global calamity.

In 2004, John Schellnhuber,[32] a distinguished science adviser at Tyndall Center for Climate Change Research in Great Britain, identified 12 global warming tipping points. If ignored long enough, any of them could initiate sudden, catastrophic changes on our planet.

As long as we continue in our bald-faced denial of climate change; as long as we maintain no personal responsibility or action; without a 13th tipping point toward stabilizing climate change—we can't hope to avoid global mayhem.

Julia Whitty, research writer, said, "The 18th century taxonomist Carolus Linnaeus named us Homo Sapiens, from the Latin 'sapiens', meaning "prudent, wise." History shows we are not born with wisdom. We evolve into it."

We better get our fannies in gear by engaging our minds! Once we engage our minds, we must activate solutions to haul our species out of this maelstrom.

In the Amazon rainforest, I witnessed the astonishing, if not sickening cutting and burning of trees at the astounding rate of 1.5 acres every second. Schellnhuber's climate model shows that a warming globe will convert the wet Amazonia forest into a savannah within this century, and "the loss of trees will render the region a net carbon dioxide producer, further accelerating global warming."

What happens next? The more we destroy the support systems developed on this planet over millions of years, the faster we destroy ocean currents that regulate climate around the globe.

"As the Atlantic warms, ice caps melt, diluting the ocean and shutting down its thermohaline circulation (THC), the oceanic rivers currently delivering the thermal equivalent of 500,000 power stations' worth of warmth to Europe," Schellnhuber said.

That in turn creates another tipping point on the Greenland Ice Sheet.

"This ice, if melted, would raise sea levels by 23 feet worldwide—not counting ice loss from the Arctic and Antarctic," Schellnhuber said.

He continued, "One tipping point affects the other in a balance as delicate as that of an acrobat's spinning dinner plates at the top of a stick. Greenland's increasing freshwater flow into the North Atlantic will impact the THC. Warm water recirculating within the central Atlantic may further rearrange airflow over the Amazon, accelerating its dry-down and tree loss, and potentially freeing as much carbon dioxide from its enormous reservoir as the 20th century's total fossil fuel output. A sudden Amazonian release would melt whatever of Greenland hadn't already melted, crashing the THC and drastically cooling Europe, in worst-case scenario, freezing it solid."

How could that happen? Please remember that glaciers covered most of North America and Europe thousands of years ago. A shift in the Humboldt and Japanese currents could warm or cool those continents into totally harsh climatic futures.

As you can see, we tinker with nature's most delicate balancing system like no other species on the planet. Like little kids, we find ourselves enamored with the operating mechanism of our new watches—wanting to see how they work. We pry off the backs and pull out the working parts. By experimentation, we place them over a candle to see if the plastic will melt. We pour Clorox on them to see if those watches can withstand chemicals. We dump oil into the mechanism to see what happens. When we try to put it all back together; uh-oh! We broke our watches with no inkling on how to fix them.

Next, we'll deal with the other tipping points that lead to deleterious global ramifications for not only humans, but for all living creatures on this planet. The second part of this report on climate change will be even more serious and sobering.

If you would like to investigate further, look for Julia Whitty in Mother Jones Magazine "*The 13ᵗʰ Tipping Point: 12 Global Disasters and 1 Powerful Antidote*" November/December 2006, page 45-51 at www.motherjones.com.

Pretending or denying climate change warming is like throwing your children into a tank of sharks and hoping the sharks suddenly become enlightened vegetarians.

CHAPTER 26: CLIMATE CHANGE TIPPING POINT

"The 13th tipping point involves humanity's return to a balanced interaction with earth that will allow our planet home to support all life in perpetuity."

FHW, environmentalist

You may argue and become indignant about whether or not climate change manifests across the globe, but it won't withstand fantasy if what the scientists tell us, is, in fact, true.

However, if you think or believe the next added 100 million Americans, as well as the next 1.5 to 2.0 billion people added to the planet by 2035, won't impact climate change, or there is no such thing as climate change, well, "Golly Andy," Goober said. "I can't figure out why it's so dad burn hot out today."

"It could be that your car's on fire, Goober," Andy said.

"Guess I better get the fire extinguisher out," Goober said.

"Not a bad idea, Goob," Andy said.

In this second part of the chapters on climate change, we discuss "tipping points" as they relate to climate change. What is a tipping point?

During a recent football season, the San Diego Chargers fell behind three touchdowns and a field goal late in the third quarter. They didn't stand a chance of winning. But a pass here, a run there, and a touchdown, then an interception led to a tipping point, where their momentum smashed their opponents and by the end of the fourth quarter, the Chargers made a last minute score that won the game.

Things ran smoothly with the planet in 1900. The planet hummed along with 76 million Americans and 1.6 billion humans. They burned whale oil, coal and wood. Then the internal combustion engine arrived on the scene. Today, with billions of cars and 6.72 billion humans burning billions of gallons of gas, fuel oil, natural gas, wood and coal, these became the factors in the "tipping point" of planetary imbalances.

History shows that scientific advances suffered scurrilous opposition by leaders who lost their religious authority. The Pope persecuted Galileo for validating that the earth revolved around the sun. Church leaders denounced Harvey for claiming blood circulated through the body instead of standing still. Newspapers branded Louis Pasteur a quack for his germ theory. Everyone 'knew' the Wright brothers couldn't get that new fangled flying machine into the air. Most conjectured that Lindbergh would not make it across the Atlantic in his "Spirit of St. Louis."

Recently, I presented my program, "*The Coming Population Crisis in America: and what you can do about it*," to a fine group of gentlemen in Denver, Colorado. No one blinked during the 30 minute program. They asked if the DVD video was available. However, at the end, two men approached and told me never to speak about global warming because they didn't want to hear about it because it was patently untrue. I felt the same gut reaction as Galileo, Pasteur, Lindbergh and Harvey probably suffered.

Again, distinguished science adviser at Tyndall Centre for Climate Change Research in the United Kingdom, John Schellnhuber cited a study whereby, "Americans fall into 'interpretive communities' that share similar world views. On one end of the spectrum, you've got naysayers who perceive climate change as nonexistent. They are predominantly white, male, politically conservative, holding pro-individualism, pro-hierarchism, anti-environmental, carrying highly religious paradigms and rely on radio as their main news source."

On the other side, alarmists scream bloody murder! In the middle, you see measured and reasoned researchers. Me? I'm an educator and messenger. You can use this information and go to the web sites provided, or you can blow it off. To a large degree, our media, which corporations control, brings you bloody killing stories from Iraq, but squashes news about 80 degree days in December in Colorado or dead zones reaching thousands of square miles in our oceans. As

I've said in the past, engage Mark Twain's famous statement about 'silent-assertion' as the shabbiest of all lies—perpetrated by politicians and the media.

For further examples, your media reports Paris Hilton's current heartbreak, but it recoils from the facts about the ozone hole expanding over Antarctica. Your complicit media in action! ABC, NBC, CBS, FOX and CNN report numerous and growing symptoms of hyper-population consequences, yet never address the causes. Their actions ensure a negative tipping point and continued subterfuge.

Since the media continues its suppression of these realities, you may investigate the web sites mentioned within this book to enlighten yourself.

For example, we hear about the connection between ozone depletion, skin cancers and cataracts, but little about the fact that increased ultraviolet radiation will destroy oxygen producing phytoplankton (80 percent of our oxygen comes from them). Without them turning sunlight into organic life, none of us would be or will be here.

Phytoplanktons create counter weights to another tipping point which is the Antarctic Circumpolar Current which circulates 34 billion gallons of water around Antarctica every second.

"When phytoplanktons die, they sink, taking the oxygen load with them to the bottom of the ocean. Global warming slows the nutrient upwelling, affecting the phytoplankton populations in the Pacific, Indian and Atlantic Oceans," Schellnhuber said. "As the oceans affect the atmosphere, the land affects the oceans. Warming will shrink the Sahara by increasing rainfall. A greener Sahara will emit less airborne desert dust to seed the oceans and feed the plankton, which in turn suppresses hurricane formations to fertilize the carbon dioxide eating trees of the Amazon. Hardly a neighborhood on earth will look the same if Africa tips."

In nature's dance, everything plays its part—whether water, land, air or fire. The more we offset the natural cycles, the greater our price. Nature rearranges herself with ruthless dexterity when we create imbalances. For example: Katrina!

"The source of Tibet's thunderstorms is the Asian monsoon, which drives moisture up the Himalayas," Schellnhuber said. "A warming climate could either weaken or strengthen the monsoon. Either effect is potentially catastrophic for more than half the world's population adapted to and reliant on the monsoon. This is another tipping point."

Garrett Hardin addresses our depredation to our planet home as the paradox of the *"Tragedy of the Commons."* As to the use of the land, i.e., our planetary resources, if an altruist and a cheater go one-on-one, the cheater wins consistently, but in the end, everyone loses. You might call the many facets—or tipping points of climate change—the growing common denominators of the *"Tragedy of the Commons."*

"In the end, this recent melting may be caused by the Antarctic Oscillation, a kind of *on/off* switch affecting pressure gradients in the Southern Hemisphere," Julia Whitty reported. "The cooler stratosphere caused by the ozone hole produces weather changes at ground level now threatening to turn Antarctica's icescape into a continent swallowing seascape. In less than 200 years, armed with fossil fuel burning, humans donned the acrobat's tights, wrested hold of the spinning plates and initiated their own wobbly circus. Nature, impassive and potent, waits to reward or punish us."

The final "tipping points" affecting our planet consist of: destroying the Amazon rainforests which cascade into changing the North Atlantic current, melting the Greenland ice sheet, destroying the ozone layer, disrupting the Circumpolar Current, changing precipitation on the Tibetan Plateau, affecting Asian monsoons, increasing methane *clathrates*, changing salinity values, offsetting El Nino, and liquefying the West Antarctic ice sheet.

What can change our fate? Who or what is the antidote to climate change? What is the 13th tipping point back toward planetary balance?

You! You and your actions! Your actions combined with everyone as knowledgeable as you! Everything you do, counts! Gandhi marched for freedom. Dr. Martin Luther King spoke of his "dream" for equality. Susan B. Anthony marched for suffrage for women. Jackie Robinson broke the color barrier in baseball. Go to www. stopglobalwarming.org for more ideas on how you can take action. You possess equivalent gifts of all your heroes. Common citizens with uncommon determination change the course of history. The next president of the United States might be another Teddy Roosevelt, Lincoln, Jefferson or Madison.

We all possess a compelling investment in the outcome.

SECTION VI: GROWTH, DOUBLE-EDGED SWORD

CHAPTER 27: CROWDING OF AMERICA

"What will we do: debate immigration policy, unborn fetus rights, whether God will come to our aid, gay rights and many other superfluous issues? The blind bus driver and his bus are about to smack the ground as it plunges off Vail Pass and the skiers are still arguing about the seating arrangements. What is truly ironic is that no one noticed the bus driver's cane and dark glasses as they got on the bus. What is also amazing is that the bus got that far, and along the way he picked up more and more passengers."

Reid, Canadian philosopher

As with all chapters of this book, they pertain to our national hyper-growth dilemma. What if, much like Katrina, we added those 100 million people in the next 14 days? Or next year? Or even five years? Any takers? You would have to be intellectually blank to choose adding that many people in the next five years or less. Or, ever!

Yet, no one utters a peep about the "Human Katrina" that will hit in three decades! It's not "if" it's going to hit folks; it's "when" it's going to hit. And, once it hits, you won't be able to hide, escape or hope it vanishes.

In history such as Hitler's rise to power, no one stood up against him. On this issue everyone waits for the first person to object. I'm standing up! I'm speaking out! I invite you to stand with me and speak out with me. Why? If not you, then whom? If not now, then when?

Crowding of America

Each year close to where I live, Rocky Mountain National Park entertains 1,000,000 people on the highest continuous highway in America, Trail Ridge Road at 12,082 feet, with a total length of 30 miles. It's jammed, crammed and hammered all summer long. People pack into campgrounds and motels with no relief. You must reserve a spot months ahead. The "wilderness experience" proves a joke! The animals run for their lives to get away from hordes of tourists. Boom-box noise ricochets off mountain sides. Fumes coalesce into killer smog in paradise.

Can you imagine 100 million people added to America when they try to visit our national parks to get away from it all?

Last summer, I visited Yellowstone, Great Smoky, Glacier, Grand Canyon, Yosemite and Arches National Parks. Each park suffers one to three million people every summer. They're so full of people; few enjoy a "wilderness" experience. Park rangers carry side arms; some rangers suffer death by criminals. Why is that? Drug pedaling villains and others toss trash and human waste everywhere in our parks. Again, imagine what an added 100 million people will do to everyone visiting those parks.

I've rafted the Grand Canyon from Lee's Ferry to Diamond Creek take-out. Today, it takes 10 to 12 years to obtain a permit to raft the Colorado River. With an added 100 million people, it will take 20 years to a lifetime to gain a permit. The more extreme our numbers the more extreme our limitations!

Back in Denver, 10 years ago, it took me 1.5 hours to go skiing and camping in the High Country. Plentiful campgrounds; safe ski slopes! Today, after Colorado added 1.3 million in a decade, campgrounds and rivers suffer overload. Ski slopes feature humans as pin balls from too many people. It takes me three to four hours to return home through gridlocked traffic. Overwhelming traffic makes Interstate 70 a gridlocked parking lot 90 miles long! Guess what? Colorado expects to add five million people.

In Denver, I-25 and 6th Avenue gridlock from 6:00 a.m. to 8:00 p.m. Twenty-five accidents disrupt motorists' lives each day from stop and go traffic. Major smash-ups happen five days a week from rush hour. Can you imagine what another two to three million more cars in Denver will do to people and their lives?

National sustainable population policy

In 1963, we housed 194 million people. As a math and science teacher in the 70s, 80s and 90s—to supplement my scant teaching salary—during the summer, I drove an eighteen-wheeler for United Van Lines through 48 states and Canada. Except for NYC, Atlanta, Chicago, Los Angeles and San Francisco, I rolled through most cities with relative ease. By 2000, we added nearly 100 million people.

It takes up to three hours to get across the Washington Bridge in New York City. Freeways jam 15 hours a day. Millions suffer "road rage" in bumper-to-bumper traffic every day of the week. Millions go to work daily, while 43,000 die in traffic annually. As the bumper-to-bumper traffic increases, billions of gallons of gasoline burn to create ever greater air pollution. So, while you stew with an ulcer and Excedrin headache in traffic, you breathe toxic air created by your own idling engine. How will another 100 million Americans improve that nightmare?

For example, Japan endures 127 million people in a landmass less than the size of California. To give you an idea where we're headed, while touring Japan, I watched baseball games within limited space. They construct four ball diamonds in a box of land all facing inward toward each other. All the outfielders mingle with outfielders from four other teams. You can't believe it unless you see it. What benefit has overpopulation given the Japanese or Chinese? Depressing doesn't begin to describe their population plights.

If you sat on a geo-stationary earth-satellite and watched the activities on earth over time, you would see drastic changes. You might notice that more and more long-haul trucking companies send their freight-trailers by railroad. Why? Gridlocked highways coast to coast! However, railroads suffer more and more crowding; tracking systems break down and staging yards bulge with too many train cars.

From your perch, you see California building one new school daily—I repeat one school daily—to keep up with the children of 1,700 people added every 24 hours. (Source: www.capsweb.org) They cannot retain qualified teachers in classrooms; thus they rank in the last five states in educational competence. Additionally, California adds 400-500 cars daily.

When I spent my childhood in LeRoy, Michigan, we owned a lot at a cost of $2,000.00 on Hogback Lake. The lake enjoyed a total of 10 cottages on it. You might see three fishing boats on the whole lake at one time. Today, close to 100 cottages swamp the lake while speedboats destroy any fishing. Each lot costs significantly more. Can you imagine what will happen with millions of people added to the USA?

This happens all over America. We clear forests, drain wetlands, pave wilderness and devour land as we explode our population. This book only touches the tip of the iceberg. In your state, you experience everything I write about. The question is: how far do you want to sink into this quicksand? How long will you remain silent? How long will you do nothing?

In every realm of everything we do in America, this population invasion will wreak havoc on the American Dream.

Trouble knocks! No! It pounds on our front door! If we continue on this current path, we face a Hobson's Choice: if we turn right, we walk over a cliff; if we turn left, we sink into quicksand.

How about changing course before Hobson's Choice kicks in?

"That's a darn good idea, Andy, to stay out of rush hour traffic in that big city of Raleigh," Barney said. "Wonder why I didn't think of it."

"Cuz you ain't got your thinkin' cap on yer dad blamed head," Goober interrupted. "You go off half cocked half the time."

"Hey, Barn, Goob," Andy said. "All ya' gotta'do is use common sense and you'll be fine."

CHAPTER 28: SUSTAINABLE GROWTH UNSUSTAINABLE

"Over-extension: our American way of life is not sustainable."

Chris Clugston, Energy Bulletin

Richard Stengel, managing editor of Time Magazine, in October 2006 wrote an essay promoting America's population growth, "We need to continue growing but in smarter more sustainable ways."

A picture of Stengel wearing a suit and tie along with a smile accompanied his essay *"Tracking America's Journey."* He looked intelligent, but his words betrayed his misunderstanding of America's population dilemma. Stengel illustrates 20th century thinking in the harsh realities of the 21st century. In other words, he's clueless as to what he's talking about. However, since he looks authoritative, millions of people subscribe to what he promotes. Since you have read this far, you know better!

Albert Einstein warned, "The problems in the world today are so enormous they cannot be solved with the level of thinking that created them."

In his essay, Stengel illustrated our glorious past population growth and projected our adding 100 million people in three decades. He said, "Unlike Japan and Europe, the U.S. is still growing at a healthy clip." He neglected to state that millions of those immigrants flee from overpopulated countries that can't feed their populations. That phenomenon fuels our population growth.

Stengel neglected to understand that a finite system cannot maintain a 'healthy' and 'sustainable' growing population ad infinitum. The two stand diametrically opposed. Stengel subscribes to antiquated 20th century thinking. He presents well, but he's totally out of touch with the consequences of what he promotes.

His kind of thinking drives California's current 37.5 million onward to 79.1 million in 40 years. Stengel's thinking adds 12 million people to Texas in 16 years.

Let's get down to brass tacks on the absurdity of unending growth and sustainability!

Dr. Albert Bartlett wrote, *"Arithmetic, Population and Energy."* You may obtain a copy of the video at the University of Colorado Boulder campus bookstore at 303-492-7599. Please call them to shatter any remaining illusion for limitless population growth you may possess. You can also order at www.cubookstore.com or tradebooks@cubookstore.com and your request will be forwarded to the correct people.

That video would cause Time Editor Richard Stengel to write a different essay on America's future. Why? He could no longer romanticize. He couldn't write glowingly about the future with an added 100 million people. He couldn't obfuscate the facts we face as a civilization headed for an unsustainable future. Dr. Bartlett writes:

The meaning of sustainability

"First, we must accept the idea that 'sustainable' has to mean 'for an unspecified long period of time'.

"Second, we must acknowledge the mathematical fact that steady growth gives very large numbers in modest periods of time. For example, a population of 10,000 people growing at 7 percent per year will become a population of 10,000,000 people in just 100 years.

"From these two statements we can see that the term 'sustainable growth' implies 'increasing endlessly', which means that the growing quantity will tend to become infinite in size. The finite size of resources, ecosystems, the environment, and the Earth, lead one to the most fundamental truth of sustainability:

"When applied to material things, the term 'sustainable growth' is an oxymoron."

Sustainability

Bartlett said, "The terms 'sustainable' and 'sustainability' burst into the global lexicon in the 1980s as the electronic news media made people increasingly aware of the growing global problems of hyper-population, drought, famine, and environmental degradation that had been the subject of *"Limits to Growth"* in the early 1970s.

"A great increase of awareness came with the publication of the report of the United Nations World Commission on Environment and Development, the Brundtland Report, which is available in bookstores under the title *"Our Common Future."*

"In graphic and heart-wrenching detail, the report places before the reader the enormous problems and suffering experienced with growing intensity every day throughout the underdeveloped world. In the foreword, before there was any definition of 'sustainable', there was the ringing call:

"What is needed now is a new era of economic growth—growth that is forceful and at the same time socially and environmentally sustainable."

"One should be struck by the fact that here is a call for 'economic growth' that is 'sustainable'. One has to ask if it is possible to have an increase in economic activity without having increases in the rates of consumption of non-renewable resources. If so, under what conditions can this happen? Are we moving toward those conditions today? What is meant by the undefined terms, 'socially sustainable' and 'environmentally sustainable'? Can we have one without the other?

"As we have seen, these two concepts of 'growth' and 'sustainability' stand in conflict with one another, yet here we see the call for both. The use of the word 'forceful' would seem to imply 'rapid', but if this is the intended meaning, it would just heighten the conflict.

"Thus sustainable development can only be pursued if population size and growth are in harmony with the changing productive potential of the ecosystem.

"One begins to feel uneasy. Population size and growth are vaguely identified as possible problem areas, but we don't know what the

Commission means by the phrase 'in harmony with...?' It can mean anything. By page 11 the Commission acknowledges that population growth is a serious problem, but then:

"The issue is not just numbers of people, but how those numbers relate to available resources."

Urgent steps are needed to limit extreme rates of population growth

Once you read or watch Dr. Bartlett's presentation, you will be more apprised of reality than Time's Editor Richard Stengel. There's no way we need to, or can add 100 million people to the United States by 2035.

I've seen Dr. Bartlett give his presentation personally. There's no dancing around his facts, figures and harsh reality check. In my world travels, I've witnessed population growth's worst outcomes. America already walks on the thin ice edge of our own demise with 306 million people.

We cannot sustain unlimited growth. We cannot break the laws of nature as to 'carrying capacity'. We cannot add 40 million more people to California and think we can provide water to drink, for crops, for animals, habitat for all other life and room to live a decent life. We cannot be THAT stupid, but, as of this writing, and in concert with Time Editor Richard Stengel, we prove Einstein a wise man, i.e., human stupidity!

"Unbeknownst to many Americans, there is overwhelming consensus among scientists that we are very close to reaching a point of no turning back on global warming, which is caused by the burning of fossil fuels. We are approaching a point at which all of the following will become unavoidable: massive desertification, rising sea level, explosive growth of insect populations, widespread habitat destruction, mass extinctions, mass migrations (including humans), the disappearance of sea life, and in all likelihood wars over drinking water that will make the wars over oil look civilized." David Swanson, ecologist.

"Exponential growth is adding one billion people to our planet every 12 years. Ninety percent of this growth stems from the developing world. The consequences are grave. Environmental destruction escalates as more people compete for water, land, clean air, food, fuel and amenities. Civil conflicts and ethnic wars roil societies as Balkanized people attempt to gain advantage through resource grabs at the expense of neighbors. Millions of the dispossessed are forced to migrate—straining the infrastructure and good will of richer nations." William B. Dickinson, author of *The Bio-Centric Imperative*.

CHAPTER 29: SELECTED GROWTH FALLACIES

"Idealism and good intentions are insufficient responses to the problems of population pressure and resource depletion."

Richard Heinberg, *Peak Everything*

How did we wander into this 'growth at all costs' trap? How could educated and rational leaders cheer unending Gross National Product? At what point does the stock market explode through the ceiling and why do we cheer its continuous growth?

That's like urging a 450 pound 5'10" man into eating a truck load of ice cream so he might 'grow' to 500 pounds. Once he reaches 500 pounds, pro-growth advocates would cheer him on to 600, 700 and 800 pounds.

How will Los Angeles survive its projected 20 million added people at even the lowest rung of Maslow's hierarchy of human needs—much less self-actualization at the top? How will California grow crops with dwindling farmland and water? Currently that state destroys 240,000 acres annually in development via concrete, asphalt, malls, roads, etc. How will drivers tolerate another 15 to 20 million cars and trucks on their already gridlocked highways of 2009? How will LAX and Ontario airports deal with more air traffic? How about the horrific air pollution generated by millions of cars, homes, schools, factories and farms?

In his book, *Collapse: How Societies Choose to Fail or Succeed*, Pulitzer Prize winning author Jared Diamond illustrated how great civilizations dismantled their own sustainability. On tiny Easter Island, humans ignored nature, cut down all the trees, built 50 ton rock monuments, then paid the price of their demise via starvation. Today, we're faced with the same dilemmas Rome, Anasazi, Mayans, Sudan, present day Rwanda and other vanished civilizations discovered.

As our numbers grow, America and most of the Western world will battle for remaining resources. Yet, the most powerful world leaders ignore this human situation on every front.

Our leaders think if we study climate change more, or ignore it further, perhaps it will vanish. If we pretend we don't have 3.5 million homeless in America, maybe they will fade into the woodwork. If we disregard our water crisis in the American West, surely, adding 30, 40 or 50 million people to Texas, California, Arizona and New Mexico won't be a problem.

Hey, I'm selling ten acre plots on the Arctic Circle for development to the first bidder at $1.00 an acre. It's a deal of a lifetime!

However, as Herman E. Daly, author of, *Steady State Economics,* said, "If you've eaten poison, you must get rid of the substances that are making you ill. Let us then, apply the stomach pump to the doctrines of economic growth that we have been forced-fed for decades."

Daly makes a point I consider critical in this debate: "We cannot have too many people alive simultaneously lest we destroy carrying capacity and thereby reduce the number of lives possible in all subsequent time periods."

Let's face it, capitalism, for all it brings us in wealth—also destroys our supporting environment at a sickening and withering pace. Capitalism's voracious drive for production and consumption knows no bounds. If allowed, it would, like a cancer cell, multiply production until it kills its host.

Daly states, "Environmental degradation is an iatrogenic disease induced by physicians (pro-growth advocates) who attempt to treat the sickness with unlimited '*wants*' by prescribing unlimited production. We do not cure a treatment-induced disease by increasing the treatment dosage."

What does that mean? You cannot keep red-lining a four cylinder engine with limited horsepower to rev it up forever in order to create more energy. You will, in the end, blow up the engine just as you would kill a horse by overloading the wagon it's pulling.

As the economist Tobin wrote, "The prevailing standard model of growth assumes that there are no limits on the expanding of supplies of nonhuman agents of production. It is a two-factor model in which production depends only on labor and reproducible capital. Land and resources, the third member of the triad, have been dropped."

In other words, ignored as a factor in GNP! That's like ignoring that our 500 pound friend has reached 600 pounds and can no longer walk because he's too fat to get onto his feet.

As Daly said, "Current economic growth has uncoupled itself from the world and has become irrelevant. Worse, it has become a blind guide."

If we bet the house on growing our population by adding 100 million and then, another 100 million, what if the hypothesis that technology will save us turns out wrong? Our bet today, more than likely in 30 years, reaps the whirlwind of a lost wager.

Please remember that as our population increases, every aspect of this book accelerates commensurately. Once manifested, there's no escape for any of us.

As Diamond illustrates in former collapsed civilizations, choices could have been made to create a sustained long term society, but weren't. Looking at China and India today, as well as having traveled through them, I see history repeating itself.

India maintains no *"National Population Policy"* while expected to grow from 1.1 billion to 1.55 billion by 2050. While China forces one child families, they still expect to grow from 1.3 billion to 1.5 billion. That's caused by a phenomenon known as 'population momentum'. Even at one child per family, their horrendous population creates a population 'overshoot' from the sheer numbers. Bangladesh, at 144 million in a landmass the size of Iowa, continues growing at breakneck speed. All three countries face crippling environmental dilemmas and human suffering on a scale we cannot imagine.

That brings me to America's predicament. Of the next added 100 million, fully 70 million will be immigrants and their children from third world countries. What they bring with them are age old rituals, mores and birth rates.

In my travel to Seattle, Washington, I talked to my cab driver from Bangladesh who had immigrated to the USA in 1990. I asked him about his country's population. He said it sickened him so much that he would not go back. I asked him why? He said his Muslim religion encouraged every woman to have as many children as possible. I noted that his country exceeded its carrying capacity and that such massive population overload would destroy any hope of 'quality of life' and decent 'standard of living.'

"Why not practice family planning?" I asked.
"It would go against Allah's wishes," he said.
"But it creates such human misery," I said.
"One must not go against Allah's wishes," he replied.
"How many kids do you have?" I asked, in frustration.
"Seven with one on the way," he said, proudly.
"How many do you expect to have?" I asked, calmly.
"That is for Allah to decide," he said.

Having lived in America for 18 years, this man had not made the connection concerning quality of life and number of children. Having been influenced by America's educational system, he was as out of touch with reality as if he remained in Bangladesh. His story stands as a glaring example why third world immigrants continue large families of six to 10 when they immigrate to first world countries.

In May 2006, via SB 2611, and in June 2007, via SB 1639, our U.S. Senate voted to change legal immigration from 1.2 million to 2.4 million annually, increase chain migration and work visas. Both Senators John McCain and Barack Obama voted for it. It failed via a grassroots uprising from Americans who understood the stakes. With the new Democratic Congress majority, it may pass, making us the receiving ground for 70 million immigrants in 30 years.

Therein lies the costly results of consciously or unconsciously adding 100 million people in 26 years—if the low estimate holds.

CHAPTER 30: TOXIC UNDERBELLY OF GROWTH

> "Mankind is looking for food not just on this planet but on others. Perhaps the time has now come to put that process into reverse. Instead of controlling the environment for the benefit of the population, maybe we should control the population to ensure the survival of our environment."
>
> Sir David Attenborough - The Life of Mammals

Have you ever heard of a "diesel death zone" in or near your neighborhood?

Have you awoken day after day to clanking machinery, wafting odors and the roar of big trucks? Have you sat in gridlock traffic coughing on toxic fumes? Have you seen black smoke belching from truck stacks? Have you noticed that brown cloud over your city?

Wade Graham in the spring 2007 issue of *"OnEarth"* wrote a brilliantly depressing article, *"Dark Side of the New Economy"* for the Natural Resources Defense Council.

The busiest seaport in the United States; San Pedro, California

"California's San Pedro Bay ports located in south Los Angeles form a vast metropolis of polluting cargo ships, trucks and locomotives— a diesel death zone," Graham said.

In this marine arena, 5,800 cargo ships unload 40 percent of all seaborne goods imported into the United States annually. Everything passes through this port including oil, cars, salt, steel, chemicals, plastics, gypsum, machinery, lumber, cotton, food and much more.

An astounding 40,000 truck trips a day move containers from docks and terminals to trains and interstates for distribution.

"Shipping volume doubled from 1990 to 2000, and doubled again by 2006," Graham said. "A conceit of the 'new economy' is that it promises freedom from smokestacks and sweatshops of the past two centuries. But this is an illusion. The new economy not only rests on the grimy pollution of the old one, but propagates, multiplies and feeds it while spreading it around the world like a pandemic."

Emissions—huge in total

Ships arriving in California burn low grade fuels that emit sulfur content at 3,000 times higher than fuel in new diesel trucks. Large cargo ships burn 'bunker' fuel that emits as much exhaust as 12,000 cars. While unloading for three days, these ships idle their engines— spewing toxic exhausts into the air 24 hours a day.

When you multiply 5,800 ships, thousands of trucks, barges, trains, homes and factories— imagine the environmental misfortune for people living in the area.

Graham said, "The twin ports emit more pollution than the top 300 industrial sources and refineries in the Los Angeles Basin combined."

Just think what happens to this port and its citizens when it receives goods for consumption by additional population.

Impact upon human beings catastrophic

"The crude machinery of 21ˢᵗ century world trade presses up against peoples' lives like a dirty storm surge," Graham said. "The smoke, smog, smell, noise and glare of lights flood the area 24 hours per day, seven days a week. Trucks are everywhere; some 15,000 rigs, heavily polluting, driving on chock-full highways while they ply local streets looking for a faster way onto jammed 710."

Jesse Marquez, local activist, said, "You see and feel the smog and smoke clouds, you breathe sudden, inexplicable miasmas of chemical stench that vanish just as suddenly, your eyes sting and your head pings. In bygone days, harbors smelled of rotting fish, creosoted pilings and a thousand dank and exotic odors of the goods that moved through them. Now the overwhelming smells come from petroleum products and their combustion."

For most innocent and unknowing Americans, diesel exhaust and all burning creates fine particles that penetrate lung tissue causing genetic and cellular tissue damage. Diesel emissions contain benzene, formaldehyde, nitrogen, sulfur oxides, arsenic, cadmium, dioxin and mercury—all cancer causing agents.

As shown in *"Chapter 8: Air Pollution"*, children suffer the greatest—with asthma and developmental damage to their growing bodies. To give you an idea of the enormity of this situation, the United States features 86 additional seaports with ships, locomotive engines and trucks spewing revolting amounts of pollution into our air, land and oceans. With our exponential population growth, this insidious assault on our environment grows worse by the day.

I lived in Colton, California for a short time. It's located at the end of a funnel cloud of pollution rolling through the San Bernardino Valley. When I jogged in the morning, it felt like I breathed air directly from the exhaust pipe of a car. By the afternoon, I felt mentally and physically exhausted. Why? Everyone breathes extremely polluted air throughout the Los Angeles area. I moved away within six months to save my lungs.

Unchecked growth; more population means more emissions

Graham said, "Total business volume expects to triple by 2020, and quadruple by 2025. Already, 50 vessels stack up at a time—waiting to unload while they idle black smoke into the skies of every port."

Graham wrote that local activists work to get the big ships hooked up to electric outlets for power while they unload their cargo shipments, but, for the most part it's a futile effort as massive growth defeats any efforts for environmental responsibility.

Laura Rodriguez, an activist to stop the pollution, pushed a bill to clean up the air, but Governor Schwarzenegger vetoed it. Big business refused to support a $30.00 per overloaded container for air quality improvement. She said, "I think that nothing we do counts."

Not against money and profits from the big boys

Please take a moment, close your eyes, and imagine how insidious pollution blankets larger metropolitan areas—coming to a city near you. Do you live in such a 'diesel death zone' or a city with massive air pollution?

Our future filled with unnecessary dangers

That's what we face fellow citizens! This unending growth paradigm bases its existence on production, consumption and waste. Rampant population growth provides the fuel that grows the wildfire of consumption resulting in pollution. The implosion becomes inevitable. As you connect the dots on this overpopulation dilemma facing not only America, but India, China, Mexico, Africa, Brazil and many other countries— can you not help but become alarmed?

As you can see by Laura Rodriguez's frustration, she felt her efforts futile. However, as this overpopulation crisis deepens and widens across America, more individuals must step forward to ignite a 'critical mass' of citizen action to stop it. Every involved American creates the vital "tipping point" for changing the future toward a viable civilization.

CHAPTER 31: TAKING IT TO THE LIMIT

"Take it to the limit, one more time."

Eagles, rock band

How far down a broken glass-filled path would you travel if you were barefoot? How long could you eat contaminated meals before you suffered food poisoning? How far could you travel in the desert without water? How long would you last driving your car 100 miles per hour in a 40 mph zone in a city? How far are you willing to take anything to the limit?

Consider today we already suffer 'over the limit' consequences without those millions yet to manifest.

- 40 percent of America's rivers and 46 percent of its lakes are too polluted for fishing and swimming in 2008.
- 6,330 species in our country stand at risk for extinction from habitat loss.
- Half of the continental United States no longer supports native vegetation since people have altered the terrain significantly with crops, farms, roads, malls, housing, airports and endless development.
- More than half the U.S. population lives within 50 miles of the coast.
- The United States continues as the third fastest growing country in the world.
- The U.S. absorbs more environmental refugees than all other nations combined. How many annually? One million legally and one million illegally add themselves to America each year.
- At 306 million in 2009, U.S. citizens burn 20 million barrels of oil daily. At 400 million, they will burn accelerating amounts per day for energy.

- The average American uses 100 gallons of water daily. It takes over 1,000 gallons of water to produce one pound of beef.
- 105,000 cows are killed daily to feed U.S. population; average of 20 million chickens killed for American consumption daily.
- Fifty percent of all wetlands in the United States have been destroyed.

With the few examples above, do you see the picture? Do you see where the United States is headed? Population drives our land use and development.

In 1997, the world exceeded its ecological carrying capacity by 39 percent which equaled 2.3 billion people. That explains why a minimum of 2.3 billion humans live in human degradation.

Population-energy expert Dell Erickson brings a more sobering aspect of carrying capacity and 'ecological footprint'. He said, "If the world's average ecological footprint in 1997 were the goal, the result would be that the world could support 3.5 billion inhabitants. The UN currently projects the world's population at 2050 to be nine to ten billion. In order for that population to maintain long term sustainability, the highest average possible living standard would be an 'ecological footprint' approximately 0.8 percent—about the living standard and footprint of today's profoundly unfortunate Ethiopia."

Would you like to live at the standard of living of present day Ethiopia when you have no other choice?

If we take this hyper-population growth to its logical conclusion, every consequence noted in this book multiplies proportionately. That's on top of the fact that another three billion people will be added to the globe. Once they arrive, they won't vanish. They will impact everything.

The most astounding aspect of what's happening to the United States stems from national denial. It pulses through our president, Congress, governors and citizens. Environmental groups like Sierra Club, Greenpeace, Nature Conservancy, Audubon, Natural Resources Defense Council, Blue Bird Society and dozens more try to 'save' animals, land and habitat. All their work proves ineffectual when they won't address overpopulation. They squander their time, money and efforts.

Once we double to 600 million—without taking action—we will grow toward one billion. India and China succeeded! Bangladesh demonstrated it could cram 144 million people into a landmass the size of Iowa. They took it to the limit; however, their societies exhibit the least human dignity, freedom and opportunity. Thus, their citizens live in a kind of living lockdown.

How far can we expand and trample our country by taking our population to the limit? What *further* spiritual disconnect do we invite by taking it to the limit when we know the limit grows deadlier by the day? Therefore, what is your responsibility to the future? Every time I turn around I hear, "I want to see my kids have a better life than I did." If we allow this massive population load addition to this country, our children don't stand a chance for a better life.

Statistics show that humans eschew change—but can we afford that stance for future generations?

CHAPTER 32: WHAT ADVANTAGES?

"Every individual and institution must think and act as a responsible trustee of Earth, seeking choices in ecology, economics and ethics that will provide a sustainable future, eliminate pollution, poverty and violence, awaken the wonder of life and foster peaceful progress in the human adventure."

John McConnell, founder of International Earth Day

Do you feel served by our president and Congress voting for unending population growth?

I asked a friend of mine, Jack Jobe in Denver, Colorado. He said, "Denial! It's the easiest emotional response for humans."

It's frustrating to be aware of America's Titanic-like course; however, you do not stand alone. Many great minds and patriots rise daily to make a stand. Many represent America's finest academics. Others represent average Americans with uncommon determination like Chris Simcox, Barbara Coe, Jim Gilchrist, Dr. Diana Hull, D.A. King, William Gheen, Roy Beck, Dr. John Tanton, John Rohe, Peter Brimelow, Terry Anderson, Tom Tancredo, Barbara Jordan, Dr. Daneen Peterson, Andy Ramirez, Lupe Moreno, Haydee Pavia, David Paxson, Richard Heinberg, James Howard Kunstler, Priscilla Espinoza, Jason Mrochek, David Durham, Dr. Albert Bartlett and thousands of others. Some speak; others write. Some create controversy while others work behind the scenes.

They possess one intention: They strive to bring about a sustainable, viable future for all Americans.

What do others say?

"The disappearance of nations would impoverish us no less than if all men had become alike with one personality, one face. Nations are the wealth of mankind, its collective personalities; the very least of them wears its own special colors and bears within itself a special facet of God's design." Alexander Solzhenitsyn

We love our families because we identify with them. We love our nations because they represent our identity. To lose one's nation is to lose one's self. Our language and our culture define and sustain us. To lose one's culture works against the human condition. One look at France, Holland, Belgium and England shows you the demise of their countries into turmoil. One week visiting Los Angeles depicts the destruction of American identity.

"The United States is not some sort of international Kleenex. I argue that borders are as essential to free societies as property rights are to free economies. When we have a clear definition of what a citizen is and what his rights and responsibilities are, we are going to maintain a civil society, an open society, a liberal democracy." Peter Brimelow, *Alien Nation*.

As you have seen in this work, we cannot survive unending growth any more than cancer cells can overwhelm their host. At some point, the cancer cells and the host die.

At some point, all of humanity and our fellow creatures face nature's response.

"You Americans live an artificially high standard of living. It's time you drop down to the poverty levels of the rest of the Third World. Immigration and population will do that to your country given enough time." Mr. Singh of Madrass, India

Do we want to grow the United States to one billion people so we can mirror the third world? What advantages will that bring our civilization?

"The modern plague of overpopulation is solvable by means we have discovered and with resources we possess. What is lacking is not the sufficient knowledge of the solution, but the universal consciousness of the gravity of the problem for billions of people who are its victims." Dr. Martin Luther King in 1966

When he accepted the Human Rights award in 1966, King understood our dilemma. Overpopulation causes greater and more profound poverty. It overburdens and devastates nature.

"Each person in the USA impacts the environment equal to a low of 10 and as high as 33 times that of a person in a third world nation. Therefore, the US population at 300 million equals in many ways a minimum impact of three billion people in environmental consequences." Royal Academy of Sciences

"Surviving like rats is not something we should bequeath to our children." Jacque Cousteau

"More than 90 percent of the increase of U.S. energy consumption from 1970 to 1990 was due to population growth. The USA continues losing more than 2.2 million acres of farmland every year to urban sprawl." Mark W. Nowak

How long can we feed ourselves by paving and cementing our precious farmland? How long can we survive by fertilizing, spraying poisons and contaminating our dwindling cropland? How can we water the crops as the underground water supplies dry up?

This book connects the dots and offers a bridge of reality to readers across the United States and the world. As you can see, we face a pretty sobering future, if—and that is a big, IF—we continue on our denial path. If you walk into a lion's den thinking he won't harm you, your denial will assure him a tasty human McMeal! If you think you can stroll 20 miles across a desert with only one quart of water on a blistering 110 degree day in Arizona, you may resemble road-kill. If we as a nation pretend that the next 100 million people won't be a problem, we stagger blindly over the edge of a cliff at the Grand Canyon. As you see, denial will debilitate this civilization.

Since we cannot depend on our leaders or Congress, we must act. The hour is late. The fat lady warms up, but she hasn't sung—yet!

CHAPTER 33: END OF THE AGE OF OIL

"As we go from this happy hydrocarbon bubble we have reached now to a renewable energy resource economy, which we must do in this century, will the "civil" part of civilization survive? As we both know there is no way that alternative energy sources can supply the amount of per capita energy we enjoy now, much less for the 9 billion expected by 2050. And energy is what keeps this game going. We are involved in a Faustian bargain—selling our economic souls for the luxurious life of the moment, but sooner or later the price has to be paid."

Walter Youngquist

In order to drive cars, boats, planes and fuel industry, Americans use 20 million barrels of oil daily while the rest of the world burns 62 million barrels. That equals 82 million barrels of oil every 24 hours!

When you multiply 365 days by 82,000,000 barrels of oil burned daily, it equals a whopping 29.9 billion barrels of oil annually.

If you remember your science, it took two billion years to produce all the oil on this planet. In other words, when oil reserves decline, we exhaust the single major energy source that drives our civilization and most other societies on this planet.

To show how much energy oil provides the U.S. annually, Michael Brownlee of www.transitionbouldercounty.org provided an astounding graph of one cubic mile of oil. That's how much oil humans burn around the planet each year! That equals to the same amount of energy provided by 52 nuclear power plants built every year for 50 years or 104 operating coal-fired electrical plants built every year for 50 years or 32,000 wind turbines built every year for 50 years—and in continuous operation—or 91,250,000 solar panels built every year for 50 years.

In other words, oil produces dramatically incredible amounts of energy that we cannot and will not be able to duplicate in the coming years.

Dr. John Tanton, publisher of *The Social Contract* at (www.thesocialcontract.com), authored, "*How Many is Twenty Million?*"

"In this age of millions, billions and trillions, it's hard to understand such numbers," Dr. Tanton said. "Twenty million is the number of barrels of oil we burn in the United States each day."

That's 42 gallons to each barrel (drum) at 30 inches tall and 20 inches in diameter, or 840,000,000 gallons burned per day. It calculates, according to Dr. Tanton's figures, to three gallons of oil per day per person in the USA. (Source: *The Social Contract*, winter 2004-05, page 151)

He said, "Suppose we took 20 million barrels and stood them side-by-side. How long a line of barrels would that make? Let's do the math: 20 inches/barrel multiplied by 20 million barrels equals 400,000,000 inches. Divide that by 12 inches/foot, and you get 33,333,333 feet. Divide that by 5,280 feet per mile, and that comes out to 6,313 miles."

Dr. Tanton computed a string of barrels, "...reaching from Seattle to Los Angeles (1,157 miles), from Los Angeles to Chicago (2,134 miles), from Chicago to Miami (1,377 miles), from Miami to New York City (1,281 miles), and from New York City to Cleveland (486 miles). Total mileage, 6,435!"

"That's how much oil we burn in the USA each day," Tanton said. "The total global consumption daily rate of 82 million would be four times this amount, or 25,000 miles—the circumference of the globe at the equator!"

Dr. Tanton asks a sobering question, "How much longer can this go on?"

The simple, unadulterated answer is: not much longer!

You may want to read, *Out of Gas: The End of the Age of Oil* by David Goodstein, physics professor at California Institute of Technology.

Another scientist, Dr. Richard C. Duncan, introduced the Olduvai Theory: *The Peak of World Oil Production and the Road to the Olduvai Gorge.*

The decline of the industrial civilization is broken into three sections:

- The Olduvai slope (<u>1979</u>–<u>1999</u>)—Energy per capita declined at 0.33 percent per year.
- The Olduvai slide (<u>2000</u>–<u>2011</u>)—Begins in 2000 with the escalating warfare in the Middle East... marks the all-time <u>peak of world oil production</u>.
- The Olduvai cliff (<u>2012</u>–<u>2030</u>)—Begins in 2012 when an epidemic of permanent blackouts spreads worldwide, i.e. first there are waves of brownouts and temporary blackouts, and then finally the electric power networks themselves expire.

We used energy from the wind, sun, animals and rivers for centuries. Two hundred years ago, we became dependent on finite oil. "We have unintentionally created a trap for ourselves," Goodstein said. "If we turn to coal and natural gas, the resultant increase in atmospheric carbon dioxide may make Earth uninhabitable. Even if human life does go on, civilization as we know it will not survive, unless we can find a way to live without fossil fuels."

World oil extraction expects to peak within a decade. Several experts predict 2010. That means no matter how many wells we drill, the earth no longer possesses 'more' oil, but, in fact, less oil. We reached "Peak Oil" in the United States in 1970. Goodstein, ever precise in his research, shows that those hoping for other fuels and substitutes to replace oil dangle their hopes by hanging from a thread over a cliff without net or parachute. To place it in a cartoon perspective, you might imagine Wiley Coyote chasing the Road Runner, but just as he grabs for his feathered friend, Wiley runs out over a cliff. He's too far out to get back! His eyes grow large; he looks below; he looks back at you; he waves bye-bye; he drops like a rock! Societies as large as humans have created, based on oil, cannot—and will not—survive.

Even if we discover huge oil deposits, it won't help much. Our expanding populations gobble them up. China stands on the verge of an industrial revolution of enormous proportions based on 1.3 billion people. They expect to place a car in every Chinaman's garage. According to James Howard Kunstler, at their current growth rate, they expect to burn 98 million barrels of oil daily by 2030! That's more than the whole world burns daily in 2009. If you look at the growth rates of 3.1 billion added to the planet by mid century, oil reserves don't stand a chance. What answer would technology give when we hit that point? Will you drive up to the gas pump? "Fill 'er up, sir!"

"With what?" the station attendant asked.

"Technology, of course!" you say.

Goodstein said, "Fossil fuels have increased the concentration of carbon dioxide from 275 parts per million before the Industrial Revolution to 370 parts per million today. Continued burning will raise it to 550 by century's end, with dire consequences as to global warming."

We examined climate change earlier in this book. However, for those that still scoff, they cannot persist with denial that nothing happens to the planet when we burn 82 million barrels of oil daily as well as millions of cords of wood, natural gas and billions of tons of coal that warm our houses. Our extreme use of energy hastens commensurate consequences.

What can we do? Goodstein said, "The best, most effective way to ameliorate the coming fuel crisis is to improve existing technologies for efficiency."

We can bury our heads in our blankets like Linus of "Peanuts" and pretend it's not coming, but the end of this "Age of Oil" threatens our civilization. It's as sure as the Exxon Valdez disaster.

"Unless we come to grips with climate change and peak oil, comments about the wonders of immigration and multiculturalism are utterly irrelevant," said Jenny Goldie.

Again, instead of growing our population via immigration, we must demand an immigration moratorium, national sustainable population policy and lead the world toward an international sustainable population policy that will allow future generations a viable planet.

To ignore our realities condemns them to the same consequences that befell the civilization and children of Easter Island. It's that simple and that brutal.

"The oil crisis may not hit until the next decade or the one following, but it will hit," said Goodstein.

CHAPTER 34: HOW AND WHY JOURNALISTS AVOID POPULATION CONNECTION

"Mark Twain's famous 'silent assertion' lives, breathes and manifests in the 21st century. The press continues to obfuscate cloud, deny, suppress and ignore America's and humanity's greatest dilemma: overpopulation."

FHW

Why do you suppose most Americans remain apathetic to our hyper-population growth? What rational person supports water shortages, climate change, and worse ramifications caused by more people?

Surprise! Most vacant-minded leaders fail to see a problem. Four years ago, I personally called U.S. Congressman Chris Cannon, (R-UT). He said, "America can easily hold one billion people."

I nearly fell out of my chair. It's almost beyond my understanding that anyone can be that 'obtuse'. [Politically correct term for—dumber than a box of gerbils!]

How do Americans live in denial? For the life of me, I don't understand.

The imminent writer T. Michael Maher said, "Recent surveys show that Americans are less concerned about population than they were 25 years ago, and they aren't connecting environmental degradation to population growth.

"Using a random sample of 150 stories about urban sprawl, endangered species and water shortages, Part I of this study shows that only about one story in ten framed population growth as a source of the problem. Further, only one story in the entire sample mentioned population stability among the realm of possible solutions. Part II presents the results of interviews with twenty-five journalists whose stories on local environmental problems omitted the causal role of population growth. It shows that journalists are aware of the controversial nature of the population issue, and prefer to avoid it if possible."

Maher continued, "In 1992 the National Academy of Sciences and the British Royal Society issued a joint statement urging world leaders to brake population growth before it is too late. That same year, 1,600 scientists (including 99 Nobel laureates) issued a statement warning all humanity that it must soon stabilize population and halt environmental destruction. That same year, world leaders ignored population growth at the largest environmental summit in history, the U.N. Conference on Environment and Development, held in Rio de Janeiro.

"Why are the American public and political leaders so indifferent about this issue that so concerns the world's leading scientists and environmentalists? Not because Americans are anti-environment: another recent Gallup Poll (Hueber, 1991), showed that 78 percent of Americans considered themselves environmentalists and 71 percent favored strong environmental protection, even at the expense of economic growth. How can Americans express strong concern about the environment, yet a diminishing concern about population growth, which many environmental experts consider the ultimate environmental problem?"

Do you see a disconnect from reality based on our history of unlimited resources, land, water and air? We continue with the myth of limitless expansion via entitlement. If we ignore it, like a child that places his/her hands over its eyes, the bad thing vanishes. If we ignore the 'monster' called too many people, since it's not harming us today, it can be ignored.

Maher explained, "Population researchers Paul and Anne Ehrlich opened their book, *The Population Explosion*,[33] with a chapter titled, "Why Isn't Everyone as Scared as We Are?" They acknowledged, "The average person, even the average scientist, seldom makes the connection between environmental problems and the population problem, and thus remains unworried." But while they noted that the evening news almost never connects population growth to environmental problems, the Ehrlichs chiefly blamed social taboos fostered by the Catholic Church and 'a colossal failure of education' for public indifference about population."

How experts frame environmental causality

Maher reported, "With specific reference to habitat loss, Sears (1956), Jackson (1981), Myers (1991), Ehrlich and Ehrlich (1990), Harrison (1992) and many others, have shown that population growth pushes people into relatively pristine, natural environments. Endangered species problems are frequently the flip side of this coin: when people convert wildlife habitat to their own habitat, they bulldoze trees, introduce chemicals, divert streams, build dams, alter the water table, and disrupt habitat in numerous other ways.

"While it is well known that environmental experts connect environmental degradation to population growth, it is less well known that land developers are equally straightforward in implicating population growth as a causal agent for turning wildlife habitat and farmland into subdivisions."

Maher said, "The search produced 1,349 water shortage stories, 1,942 urban sprawl stories, and 6,001 endangered species stories. To be considered for coding, the story had to describe a population-driven environmental conflict.

"Of the 150-article sample, 16 (less than 11 percent) mentioned population growth as a cause of the environmental problem described in the story. Population growth appeared in eight urban sprawl stories, seven water shortage stories, and one story on endangered species.

"Although many scientific groups, environmental scientists and even land development experts agree that population growth is a basic cause of environmental change, media framing diverges widely from expert framing. Just over 10 percent of a Lexis-Nexis sample of environmental news stories links human population growth to the environmental problems it affects.

"Even more significantly, only one story in a sample of 150 presents the view that limiting population growth might be a solution to environmental problems. Such stories effectively tell the reader: population growth affects environmental degradation, but population stability is too outlandish even to be mentioned as a policy option."

Ignoring that a stable population might be a long-term solution to environmental problems, news stories instead direct the public's attention to palliative solutions: build new dams to supply water, zone to prevent urban sprawl, set aside land for endangered species. In my state of Colorado, the governor built more dams, created more light rail and built more lanes on Interstate 25. Those actions, in fact, cause more growth rather than alleviate the problem. Result: demographers predict Colorado to add six million by mid century.

"In thousands of communities across America, population growth wreaks changes: a mobile home park displaces an orchard, a farmer loses his water rights to a city hundreds of miles away, an endangered reptile's last known habitat is threatened by a subdivision," Maher said. "These and countless other population-influenced disruptions reduce wildlife habitat, rural solitude, water availability, and many other environmental qualities.

"But this study shows that only one news story in ten connects these events to population growth."

In plain English, those writers entrusted with informing the American public, pass the buck, which verifies the genius of Mark Twain's 'silent assertion'. Again, those writers sustain Einstein's adage when he said, "There are two things infinite: the universe and human stupidity."

CHAPTER 35: UTTER INCOMPETENCE OF OUR LEADERS

"Cautious, careful people, always casting about to preserve their reputation and social standing, never can bring about a reform. Those who are really in earnest must be willing to be anything or nothing in the world's estimation, and publicly and privately, in season and out, avow their sympathy with despised and persecuted ideas and their advocates, and bear the consequences."

Susan B. Anthony, women's suffrage

As you read through this information, you may discover a new level of education, awareness and intellectual sobriety. We as a species and as a civilization stand at the edge of the abyss.

Astoundingly enough, our world and national leaders look into it with intrepid stupidity. Our church leaders cannot bring themselves out of their religious stupors long enough to face facts. Regular citizens stand in denial or awake completely clueless each day.

As long as we cannot experience what we see on television around the world, it's easy to live in denial or refutation of our accelerating dilemma.

For example, everyone 'watched' what happened to New Orleans during Hurricane Katrina, but 98 percent of Americans did not experience it or its current ongoing aftermath four years later. The horror of 9/11 affected 3,000 people and several thousand loved ones—with death. The clean-up crews agonized through it, but for the most part, the country didn't lose much sleep over the World Trade Towers collapsing.

Let's revisit Eleanor Roosevelt's wise words: "We must prevent human tragedy rather than run around trying to save ourselves after an event has already occurred. Unfortunately, history clearly shows that we arrive at catastrophe by failing to meet the situation, by failing to act when we should have acted. The opportunity passes us by and the next disaster is always more difficult and compounded than the last one."

Do you see any national or world leaders addressing overpopulation? Which U.S. governors address it? Have any leaders offered solutions? The harsh answer: no!

Yes, you hear about Al Gore's crusade concerning climate change. However, he does not mention population stabilization as a remedy when population increases directly affect climate change. You watch church organizations like "Save the Children" show horrific video projections of millions of starving children in Africa and other parts of the world. They invite you to send money to save starving children. But, they won't provide birth control. That means the more children they save beget millions more that will starve at a later date. You see the Pope pray for food and assistance for starving masses, but he won't advocate for birth control. You see our president advocating for more oil drilling, but not for conservation.

Everything you've read in this book occurs in greater degrees daily as the world population grows by 77 million annually. It's not getting better; it's becoming much worse.

So why do our leaders fail us?

They fail because we fail to elect visionary leaders. We respond to emotions rather than conditions. We await this "Human Katrina" rather than work to circumvent it.

They and we fail because our culture expects unlimited and unending access to expansion, growth and resources—The American Dream. Ruthless capitalism pleases many in the short term, but decimates our planet home in the long term—as I've said, the ultimate Faustian bargain.

Our leaders and we fail because not enough people understand what I hope you understand by reading this book.

Iain Orr said, "The world has three environmental crises—overpopulation, loss of biodiversity and scarcity of water resources—which would still be there if a technological fix to climate change were found tomorrow."

CHAPTER 36: NEW DIRECTION ON HUMAN POPULATION

"We cannot rein in industry if we cannot reach mutual understanding and mutual agreement based on a world-centric moral perspective concerning the global commons. And we reach the world-centric moral perspective through a difficult and laborious process of interior growth and transcendence."

Ken Wilber

"This planet ain't big enough for the 6,700,000,000 humans," said Chris Rapley of the Belfast Telegraph in the United Kingdom.

Again, if you visit www.populationmedia.org or www.worldpopulationbalance.org, you may see how fast human numbers explode across the planet on a minute by minute basis. It's pretty unnerving to see it in front of your eyes. For me, it's beyond unnerving because I've seen the human debris of those numbers on my world bicycle travels. The human race stands up to its eyebrows in trouble.

"Behind the climate crisis lurks a global issue that no one wants to tackle: do we need radical plans to reduce the world's population?" Rapley said. "What do the following have in common: the carbon dioxide content of the atmosphere, Earth's average temperature and the size of the human population?

"Answer: each was, for a long period of Earth's history, held in a state of equilibrium. Whether it's the burning of fossil fuels versus the rate at which plants absorb carbon, or the heat absorbed from sunshine versus the heat reflected back into space, or global birth rates versus death rates—each is governed by the difference between an inflow and an outflow, and even small imbalances can have large effects. At present, all of these three are out of balance as a result of human actions. And each of these imbalances is creating a major problem."

Rapley continued, "Second question: how do these three differ? Answer: human carbon emissions and climate change are big issues at the top of the news agenda. And rightly so, since they pose a substantial threat. But population growth is almost entirely ignored. Which is odd, since it is at the root of the environmental crisis, and it represents a danger to health and socio-economic development.

"The statistics are quite remarkable. For most of the two million years of human history, the population was less than a quarter of a million. The advent of agriculture led to a sustained increase, but it took thousands of years, until 1800, before the planet was host to a billion humans. Since then growth has accelerated - we hit two billion in 1930, three billion in 1960, four billion in 1975, five billion in 1987 and six billion in 1999. Today's grand total is estimated to be 6.7 billion, with a growth rate of 80 million each year.

"To what can we attribute such a dramatic rise? Impressive increases in the food supply have played a part, but the underlying driver has been the shift from an "organic" society, in which energy was drawn from the wind, water, beasts of burden (including humans) and wood, to a fossil fuel-based world in which most of our energy is obtained by burning coal, oil and gas. This transition has promoted the changes in quality of life associated with modern technology, especially the major advances in hygiene and medicine. Although unevenly distributed, these bounties have seen life expectancy double and a corresponding reduction in mortality rates.

"But success in reducing mortality has not been matched by a lowering of the birth rate—and this has resulted in the dramatic increase in the human stock. As noted by Malthus, who at the end of the 18th century was the first to foresee the problems of population growth, such growth can accelerate rapidly since every individual has the capacity to produce many offspring, each of whom can in turn produce many more, and the process will only cease when something happens to bring birth rate and death rate once more into balance.

"In fact, the overall growth rate of the world's population hit a peak of about two per cent per year in the late sixties and has since fallen to 1.3 per cent. Although the timing and magnitude of the changes have been different in different parts of the world, the pattern has followed the so-called 'demographic transition'. Initially both mortality and birth rates are high, with the population stable. As living standards rise and health conditions improve, the mortality rate decreases. The resulting difference between the numbers of births and deaths causes the population to increase. Eventually, the birth rate decreases until a new balance is achieved and the population again stabilizes, but at a new and higher level.

"Demographers offer two possible explanations for the decline in birth rate, suggesting that it is an inherent tendency of societies to find equilibrium between births and deaths, with the lag simply being the time taken for the change in mortality rate to be recognized. Alternatively, it is attributed to the same general driving forces that caused the decline in mortality, such as improvements in medical practice and technology, in this case birth control.

"So where do we stand today? Worldwide, the birth rate is about six per second, and the death rate stands at three per second. UN figures foresee numbers leveling out at a point when we have between nine and 10 billion humans by 2050 –that's roughly a 50 percent increase on today's figure.

"This is not comforting news. Even at current levels, the World Health Organization reports that more than three billion people are malnourished. And although food availability continues to grow, per capita grain availability has been declining since the eighties. Technology may continue to push back the limits, but 50 percent of plants and animals are already harvested for our use, creating a huge impact on our partner species and the world's ecosystems. And it is the airborne waste from our energy production that is driving climate change.

"Yet, even at a geo-political level, population control is rarely discussed. Today, however, marks the publication of a new report on population by the United Nations Environment Program. Perhaps this could be the spur we need.

"If debate is started, some will say that we need to stop the world's population booming, and to do so most urgently where the birth rates are highest - the developing world. Others may argue that it is in the developed world, where the impact of individuals is highest, that we should concentrate efforts. A third view is to ignore population and to focus on human consumption.

"Programs that seek actively to reduce birth rates find that three conditions must be met. First, birth control must be within the scope of conscious choice. Second, there must be real advantages to having a smaller family—if no provision is made for peoples' old age; the incentive is to have more children. Third, the means of control must be available—but also to be socially acceptable, and combined with education and emancipation of girls and women.

"The human multitude has become a force at the planetary scale. Collectively, our exploitation of the world's resources has already reached a level that, according to the World Wildlife Fund, could only be sustained on a planet 25 per cent larger than our own.

"Confronted with this state of affairs, there is much discussion about how to respond to human impacts on the planet and especially on how to reduce human carbon emissions. Various technical fixes and changes in behavior are proposed, the former generally having price tags in the order of trillions of dollars. Spread over several decades, these are arguably affordable, and to be preferred to the environmental damage and economic collapse which may otherwise occur.

"But by avoiding a fraction of the projected population increase, the emissions savings could be significant and would be at a cost, based on UN experience of reproductive health programs, that would be as little as one-thousandth of the technological fixes. The reality is that while the footprint of each individual cannot be reduced to zero, the absence of an individual does do so.

"Although I'm now the director of the British Antarctic Survey, I was previously executive director of the International Geosphere-Biosphere program, looking at the chemistry and biology of how Earth works as a system. About 18 months ago, I wrote an article for the BBC Green Room website in which I raised the issues: "So if we believe that the size of the human footprint is a serious problem (and there is much evidence for this) then a rational view would be that along with a raft of measures to reduce the footprint per person, the issue of population management must be addressed.

"In practice, of course, it is a bombshell of a topic, with profound and emotive issues of ethics, morality, equity and practicability. So controversial is the subject that it has become the Cinderella of the great sustainability debate—rarely visible in public, or even in private. In interdisciplinary meetings addressing how the planet functions as an integrated whole, demographers and population specialists are usually notable by their absence. Rare, indeed, are the opportunities for religious leaders, philosophers, moralists, policy-makers, politicians and the global public to debate the trajectory of the world's human population in the context of its stress on the Earth system, and to decide what might be done.

"The response from around the world was strong and positive—along the lines of 'at last, this issue has been raised'. But after that initial burst of enthusiasm, I find that little has changed. This is a pity, since as time passes, so our ability to leave the world in a better state is reduced. Today's report from the UN provides an opportunity to raise the debate once again. For the sake of future generations, I hope that others will this time take up the challenge."

How can an international journalist like Chris Rapley write about our human dilemma while our national leaders avoid it? If our leaders continue failing to raise this issue into the national and international spotlight, we face definite consequences as related by Garrett Hardin.

Garrett Hardin's Three Laws of Human Ecology

Garrett Hardin[34] presented these three laws of human ecology, which are fundamental, and need to be known and recognized by all that would speak of sustainability.

First Law: "We can never do merely one thing."

In other words, we all connect into the web of life. I quote from one of my favorite authors, John Muir who said, "When you pick up a rock, you realize it's hitched to the universe."

Second Law: "There's no 'away' to throw to."

This law illustrates that when a person 'throws something away', it goes somewhere. That means acid rain, chemicals, trash and pollution of any kind will affect something else in this limited biosphere.

Third Law: The impact (I) of any group or nation on the environment is represented qualitatively by the relation:

$$I = P\,A\,T$$

Here P is the size of the population, A is the per-capita affluence, measured by per-capita annual consumption, and T is a measure of the damage done by the technologies that are used in supplying the consumption. Hardin attributes this law to Ehrlich and Holdren. (Ehrlich and Holdren 1971)

As Rapley so aptly said it, we continue ignoring the obvious—at our peril. My best guess? The United States staggers along the same path as China, India, Bangladesh, Mexico, Africa and the rest of them until our misery exceeds our ability to continue growing.

Unfortunately, by that time, it will not be pretty. In fact, extreme ugliness awaits us.

Jacques-Yves Cousteau said, "We must alert and organize the world's people to pressure world leaders to take specific steps to solve the two root causes of our environmental crisis—exploding population growth and wasteful consumption of irreplaceable resources. Over-consumption and over-population underlie every environmental problem we face today."

CHAPTER 37: QUALITY OF LIFE

"Democracy cannot survive overpopulation. Human dignity cannot survive it. Convenience and decency cannot survive it. As you put more and more people into the world, the value of life not only declines, it disappears. It doesn't matter if someone dies. The more people there are, the less one individual matters."

Isaac Asimov

Thomas Jefferson proposed that every American enjoy, "Life, liberty and the pursuit of happiness." Back in 1776, that meant food, shelter, a rich and rewarding family-life, spiritual awakening, employment and creative expression.

High-speed, high-stress life—is that what you want?

In the 21st century, another phrase becomes more important in our high speed— high stress lives. *Quality of life* surfaced in the American lexicon in the last twenty years. Why? Because we grew too much, moved too fast and suffered accelerating consequences.

For those of you who experienced the 1950s and 60s, no one ever heard about grid-lock traffic, air pollution, species extinction, global warming, zip codes, overpopulation, cell phones, computers or drugs. Most guys knew every make, model and year of our cars. Choices included Chevy, Plymouth, Ford, Chrysler and some strange little Japanese import called a Datsun.

In 1965, Elvis Presley drove Cadillac convertibles while John Wayne assured us that the 'good guys' always wore white hats. Jimmy Durante clowned on TV while Frank Sinatra crooned in Vegas. Los Angeles enjoyed Sunset Strip, Route 66 and Hollywood. California sported 16 million people.

What we face!

Forty years later, California sports 37.5 million high stressed, high speed, road raging and cell-phone talking citizens. They endure gridlocked traffic, 1.5 hour commutes, air-polluted and gang-infested towns and cities. That's for starters! California expects an added 40 million people by 2050.

Would anyone say that the quality of life in Los Angeles, San Francisco, Chicago, Atlanta, New York City, Detroit, Miami, Houston, and our nation's capital— or any other multiple million populated city— measures up to something envisioned by our founders as reasonable and appropriate?

I spent four days in Washington, DC at the annual www.fairus.org *"Hold Their Feet to the Fire"* conference where 35 national radio hosts and hundreds of their esteemed guests spoke to millions of listeners concerning unchecked immigration. If I could hang a literary handle on the East Coast, I'd say, "too many people, too many cars, too much noise, too many accidents, unbelievable gridlocked traffic and stinky"— just for starters!

In truth, DC took my breath away

I am astounded that people hit the expressways at 4:30 a.m., daily, to beat the rush hour. I jumped on the Metro Subway at Shady Grove 30 miles out of Washington, DC. By the time I reached Union Station, I felt like a sardine crammed into a tin.

Unpleasant, unhealthy, obnoxious and insufferable! As I stepped off the subway, little old ladies tried to run me over on their rush to the escalator. I felt like a cork being swept away in a human ocean of people. Every metropolitan arena features maddening crowds and endless congestion.

The traffic within and around our cities grows to crisis gridlock levels. What's the latest plan to alleviate the beltways around our cities? Engineers plan to build second beltways around the first beltways— some are already completed!

The future will bring more overcrowding, unless

Where does hyper-population growth lead us? What about quality of life? What about peace of mind? First, a reminder— with thanks to Wikipedia:

> Overpopulation is the condition of any organism's numbers exceeding the carrying capacity of its ecological niche. In common parlance, the term usually refers to the relationship between the human population and its environment, the Earth.

> Overpopulation is not simply a function of the size or density of the population, but rather the number of individuals compared to the resources (for example, food production or water resources) and 'personal space' needed for healthy survival or well-being.

Are these simple examples good things?

Stop for a moment and close your eyes as you consider our current dilemma. Have you noticed the little advantages we lose as we overpopulate?

- You must dial all 10 digits for local phone calls today because of our massive population overload in our cities— some regions feature five area codes.
- Commuters added 20 minutes to commute times in the past 15 years. They can expect to add another 20 minutes in the coming 10 years. It's not uncommon in many states for 1.5 hour commute times.
- In some regions you are already expected to make reservations by lottery for our national parks— within 20 years, we will all be doing that because too many people want to visit those limited spaces.
- As highways overload with another 100 million people, your chances of making it to and from your destination erode dramatically— with more risk. Currently, 43,000 people

die from traffic accidents every year, and tens of thousands suffer injuries to one degree or another in "accidents." Those numbers grow exponentially as we add 100 million people.

- The "quality-time" that nourishes each of us diminishes; we suffer that loss.
- Family doctors report that 75 percent of diagnoses are stress related.

You can enumerate another 50 examples of the loss of quality of life from your neck of the woods.

Change can be managed, and strategies chosen sensibly

Wikipedia continues to illuminate as we allow the magnitude and depth of the personal impacts to sink in here:

"In the context of human societies, overpopulation occurs when the population density is so great as to actually cause an impaired quality of life, serious environmental degradation, and/or long-term shortages of essential goods and services. This is the definition used by popular dictionaries such Merriam-Webster.

"Overpopulation is not merely an imbalance between the number of individuals compared to the resources needed for survival, or a ratio of population over resources, or a function of the number or density of individuals, compared to the resources (i.e. food production) they need to survive.

"It is rather a situation of shortage of resources and elbow-room that must be caused by population, and not by other factors. This is because such an imbalance may be caused by any number of other factors such as bad governance, war, corruption or endemic poverty.

"People feel each human-crowding factor as an agitation, a gnawing and bothersome affect that is difficult to describe— until you venture to a pristine natural area, and begin to breathe deeply. Then, you remember what you really want, and don't."

Please ask yourself; why choose to drift aimlessly into traumatic overcrowding? America invites its hyper-population growth by absorbing millions of immigrants annually.

Strategic action and planning work!

It may startle you to learn a basic reality; China, India and Mexico already manifest what we move toward. They planned no strategy for gentle population balancing, and neither do we.

What are we thinking? In this fragile Republic of the United States, we participate in a grand experiment. Until recently, the ride proved fun. Of late, the elbow room to play diminishes—along with our creative-competitive energies.

How will you turn the horses and halt our unchecked population growth? Is your 'peace-of-mind' and optimized quality-of-life worth the effort? *Do you owe that healthy legacy to your children?*

In the end, an accelerating Gross National Product along with an exploding stock market cannot improve quality of life at this stage of our civilization. Instead, we must move toward quality of place, quality of living and quality of surroundings. How about a happy existence with a butterfly-filled open meadow in which to walk or a pristine shoreline on which to squeeze sand through our toes, or perhaps holding hands with your lover as you walk under a crimson sky at sunset? Let's move toward stress-free lives that make contributions to a civilized and sustainable world.

CHAPTER 38: HEADLINES OF OUR TIMES

"Many days we shake our heads because the news is so bad, we ignore it, and for some reason, we go on living and don't think much about it. However, that won't work for our children's outlook. We must take action in order to change the present and the future. This is a matter of integrity and personal responsibility."

FHW

Perhaps, after reading this book, you wish you didn't know what you know right now! You might be distressed or angry. You might think everything in this book cannot be happening or that it's all conjecture on the part of the author.

Therefore, I researched far and wide for news stories from around the country and planet that directly validate in a variety of ways what I have written and what I have witnessed in my world travels.

The following journalists report the facts that impact every human, plant and animal in our country and on this planet.

Again, as our numbers accelerate, everything you read in this chapter worsens all over the globe. We either stop our abuse of our planet home, or it will stop us.

News headlines that cannot be ignored—hold on to your hat

Who would have predicted a century ago that the richest civilization in history would be made up of polluted tracts of suburban development dominated by the private automobile, shopping malls, and a throwaway economy? Surely, this is not the ultimate fulfillment of our destiny.

Alan Durning, *How Much Is Enough?* 1992

Over 24 years, the government and private research agencies dumped almost 48,000 55-gallon drums of radioactive waste just a few miles west of the Golden Gate Bridge. That waste now is leaking into the Gulf of the Farallones National Marine Sanctuary—and no one knows how much contamination it is causing in seafood. Most are so corroded that moving them would spread more radioactive waste. It was dumped between 1946 and 1970 at what then was called the Farallon Islands Nuclear Waste Site.

> Colleen Valles, "Bay Area May Be in
> Hot Water Over Dumping,"
> *LA Times.com*, 17 Feb 02

The safe disposal of sewage sludge is an enormous task. American sewage treatment plants produce 11.6 billion pounds of sewage sludge each year. More than a third is spread on farmland or otherwise mixed into soils. In addition to being "human manure," sewage sludge can contain toxic chemicals, heavy metals and pathogens.

> Michael Vatalaro, *"EPA Intimidates
> Sludge Critics, Congress Told"*
> *Environmental News Service* 22 Mar 00

A dense blanket of pollution, dubbed the "Asian Brown Cloud," is hovering over South Asia, with scientists warning it could kill millions of people in the region, and pose a global threat. In the biggest-ever study of the phenomenon, 200 scientists warned that the cloud, estimated to be two miles (three kilometers) thick, is responsible for hundreds of thousands of deaths a year from respiratory disease. The potent haze lying over the entire Indian subcontinent—from Sri Lanka to Afghanistan—has led to some erratic weather, sparking flooding in Bangladesh, Nepal and northeastern India, but drought in Pakistan and northwestern India.

> Marianne Bray, *"Asian Brown Cloud poses global threat"*
> *CNN.com 12 Aug 02*

Every one of you sitting here today is carrying at least 500 measurable chemicals in your body that were never in anybody's body before the 1920s. We have dusted the globe with man-made chemicals that can undermine the development of the brain and behavior, and the endocrine, immune and reproductive systems, vital systems that assure perpetuity. You are not exposed to one chemical at a time, but a complex mixture of chemicals that changes day by day, hour by hour, depending on where you are and the environment you are in. In the United States alone, it is estimated that over 72,000 different chemicals are used regularly.

> Theo Colburn, "*Speech at the State of the Word Forum*"
> *San Francisco, 3 Oct 96*

Without requiring lab tests to determine their safety, the U.S. government has approved thousands of chemicals for use in such products as sofa cushions, soaps, paints and baby bottles. On average, two more chemicals are approved every day. The result: Consumers are unwittingly part of a vast, uncontrolled lab experiment. "We're treating [people] worse than lab rats," said Karen Florini, a lawyer with the nonprofit group Environmental Defense.

> Tom Avril, "*U.S. chemical regulation leaves much unknown*"
> *Philadelphia Inquirer, 4 Nov 03*

More than eight million pounds of persistent toxic metals (like lead and mercury) were released into our waterways (in 1997), an increase of more than 50 percent from the previous year and the largest amount since at least 1992. Nearly 900,000 pounds of reproductive toxins like toluene were released into U.S. waterways, an increase of 60 percent from the previous year and the largest amount released since at least 1992. The parent corporations with the greatest amounts of toxic pollution to waterways were Armco Inc., PCS Nitrogen Fertilizer LP, BASF Corporation, E. I. Du Pont De Nemours & Co., and Vicksburg Chemical Co.

> Cat Lazaroff, "*Polluters Sully US Waters
> Despite Federal Regulations*"
> *Environment News Service, 17 Feb 00*

By the time the Raccoon River winds through the western hills here, passing corn fields and livestock pens before reaching Des Moines miles to the east, it is so polluted the city has to put it through a special nutrient filter to meet government standards for drinking water. Across the country, metropolitan water agencies are battling increasing pollution from the countryside. The river pollution is spreading and helping to cause dead zones in the open seas. A recent study by the Pew Oceans Commission, an independent group examining government policies, called huge livestock feedlots and farm fertilizer runoff among the fastest-growing sources of pollution in oceans thousands of miles away.

Elizabeth Becker, *"Big farms making a mess of U.S. Waters"*
The New York Times, 10 Feb 02

Hundreds of millions of tiny plastic pellets, or nurdles—the raw materials for the plastic industry—are lost or spilled every year, working their way into the sea. These pollutants act as chemical sponges attracting man-made chemicals such as hydrocarbons and the pesticide DDT. They then enter the food chain. "What goes into the ocean goes into these animals and onto your dinner plate. It's that simple," said Dr Eriksen.

Kathy Marks & Daniel Howden, *"The World's Dump"*
The Independent UK, 6 Feb 08

Dust and soot in the air contribute to between 20 and 200 early deaths each day in America's biggest cities, according to the largest coast-to-coast scientific study of the problem. Ill health from particulates, tiny specks smaller than the width of a human hair, is spread across 20 of the largest cities in the United States—including Los Angeles, Santa Ana-Anaheim, San Bernardino and three other California areas—which are inhabited by about 50 million people. The researchers found strong evidence that dust and soot particles, not other factors suggested by industry, appear to be causing the harmful effects.

Gary Polakovic, *"Study Links Deaths*
to Airborne Particles"
Los Angeles Times 14 Dec 00

Between November of 1999 and December of 2000, EPA filed lawsuits against nine power companies for expanding their plants without obtaining New Source Review permits and the up-to-date pollution controls required by law. The companies named in our lawsuits emit an incredible 5 million tons of sulfur dioxide every year (a quarter of the emissions in the entire country) as well as 2 million tons of nitrogen oxide. Data supplied to the Senate Environment Committee by EPA last year estimate the annual health bill from 7 million tons of SO2 and NO2: more than 10,800 premature deaths; at least 5,400 incidents of chronic bronchitis; more than 5,100 hospital emergency visits; and over 1.5 million lost work days. Add to that severe damage to our natural resources, as acid rain attacks soils and plants and deposits nitrogen in the Chesapeake Bay and other critical bodies of water.

> Eric V. Schaeffer, Director of U.S. Environmental
> Protection Agency's Office of Regulatory
> Enforcement, Resignation Letter, 27 Feb 02

The Environmental Protection Agency believes that about 630,000 of the roughly 4 million babies born annually in the United States — twice as many as previously thought — may be exposed to dangerous levels of mercury in the womb, according to an analysis released Thursday. The primary source of newborns' exposure to mercury is the fish and shellfish their mothers eat. Mercury in children can impair motor functions, learning capacity, vision and memory, and can cause a variety of other symptoms related to neurological damage.

> Elizabeth Shogren, *"Estimate of Fetuses
> Exposed to High Mercury Doubles"*
> *LATimes.com, 6 Feb 04*

We are told that we cannot afford clean air and water and health for our children. Yet in the first few months of 2001, you and I spent over $2 billion buying videos. Brides-to-be will spend over $35 billion on weddings this year, and Americans will spend a staggering $550 billion on gambling. Corporations will spend untold billions on advertising.

> Jackie Alan Giuliano, *"Earth Day 2001 -
> A Celebration or a Wake?"*
> *Environmental News Service, 20 Apr 01*

America's cities, blanketed with smog and climate-altering carbon dioxide, have become cradles of ill health and are fostering an epidemic of asthma, according to a report yesterday from a leading group of Harvard University researchers and the American Public Health Association. Particularly hard hit are preschool-aged children, whose rate of asthma rose by 160 per cent between 1980 and 1994 (more than twice the national average). As well, particulates—or small bits—from burned diesel fuel attach themselves to mold and pollen, which in turn is delivered deep into human lung sacs. A measure of the impact is that a quarter of the children living in Harlem are asthmatic, and they are concentrated along bus routes, the researchers said.

Alanna Mitchell, *"Global warming*
linked to high asthma rates"
The Globe and Mail 30 Apr 04

The dead zone this summer reached 8,500 square miles, about as big as Massachusetts, to become the largest mass of oxygen-starved water ever recorded in the Gulf of Mexico. Crab carcasses lie covered in a bacterial mat as if spray painted white. In pockets where oxygen is totally depleted, the surface may appear clear, if a bit too glassy, while bottom waters faintly smell of rotten eggs. "Call it the Berlin Wall of the gulf," said former Louisiana shrimper Donald Lirette, "because life can't cross it from either side."

Rick Montgomery, *Knight Ridder News Service,*
"Sea suffocates in 'dead zone"
St. Paul Pioneer Press, 29 Oct 02

A first-of-its-kind study of Iowa's 132 lakes shows they are among the most fertilizer-polluted waters on Earth. "We suspected Iowa has some of the most nutrient-rich water in the world, and this proves it," Iowa State University researcher John Downing said Monday. Downing's conclusion is based on three rounds of samples from each of Iowa's 132 lakes, all taken last summer. The samples show heavy concentrations of nitrates and phosphorus, two common ingredients in farm and yard runoff.

Perry Beeman, *"Iowa's lakes among filthiest in the world"*
The Des Moines Register, 6 Mar 01

For nearly 40 years, while producing the now-banned industrial coolants known as PCBs at a local factory, Monsanto Co. routinely

182

discharged toxic waste into a west Anniston creek [Alabama] and dumped millions of pounds of PCBs into oozing open-pit landfills. And thousands of pages of Monsanto documents—many emblazoned with warnings such as "CONFIDENTIAL: Read and Destroy" –show that for decades; the corporate giant concealed what it did and what it knew. In 1966, Monsanto managers discovered that fish submerged in that creek turned belly-up within 10 seconds, spurting blood and shedding skin as if dunked into boiling water. In 1969, they found fish in another creek with 7,500 times the legal PCB levels. They ordered its conclusion changed from "slightly tumorigenic" to "does not appear to be carcinogenic."

Michael Grunwald, *"Monsanto Hid
Decades of Pollution"*
Washington Post, 1 Jan 02

During those years, St. Louis-based Monsanto flushed tens of thousands of pounds of PCB's and other toxic wastes into Snow Creek each year. More than 45 tons of PCB's, a highly efficient industrial insulator, were discharged in 1969 alone. Monsanto also deposited millions of pounds of PCB's in a hillside landfill just above the plant.

"This is by far the most contaminated community — as indicated by the levels in their blood — that I've ever encountered," Dr. Nisbet said.

"We would all rather live in a pristine world," said Jere White, a lawyer for Monsanto and Solutia, in his opening argument two weeks ago. "We are all going to be exposed to things on a daily basis. Our bodies can deal with it."

Kevin Sack, *"PCB pollution suits have day in court
in Alabama"* New York Times, 27 Jan 02

A chemical widely found in food packaging and other plastics may cause severe genetic defects in embryos, at levels people are commonly exposed to, according to a scientific study published today. Laboratory experiments by geneticists at Case Western Reserve University in Ohio showed that bisphenol 'A' disrupts the way that chromosomes

183

align to produce the eggs of mice, leading to aneuploidy, which is the main cause of miscarriages and Down's syndrome in humans. Scientists say the study is the first to show that exposure to a small amount of an environmental contaminant that mimics the hormone estrogen disrupts the growth of embryos, killing them or leading to genetically abnormal offspring.

Marla Cone, *"Study Links Plastics to Embryo Ills"*
Los Angeles Times, 1 Apr 03

"The Environmental Protection Agency concluded yesterday that long-term exposure to exhaust from diesel engines likely causes lung cancer in humans and triggers a variety of other lung and respiratory illnesses. The study, the culmination of decades of research, highlights the health problems posed by the complex mix of gases and fine particles emitted by heavy-duty diesel engines operating on the nation's highways, farms and construction sites. "Overall, the evidence for a potential cancer hazard to humans resulting from chronic inhalation exposure to [diesel emissions] is persuasive," the report states.

Eric Pianin, *"EPA Links Lung Cancer,*
Diesel Exhaust"
The Washington Post, 4 Sep 02

Exposure to the pesticide methyl bromide and six other pesticides have been linked with an increased risk of prostate cancer among pesticide applicators in North Carolina and Iowa, U.S. government scientists reported Thursday. Methyl bromide is a fumigant gas used to protect crops from pests in the soil and to fumigate grain bins and other agricultural storage areas. Prostate cancer risks were two to four times higher among pesticide applicators than among men who were not exposed to methyl bromide.

Environment News Service,
"Methyl Bromide Exposure Raises
Prostate Cancer Risk" 2 May 03

Last year approximately 400,000,000 gallons of chemical termiticides were pumped onto American soil. That's enough chemical to fill 80,000 semi-tanker trucks.

"Current termite control practices are hazardous for new homeowners, who are not even required to be notified of toxic chemical use (soil poisons)," said Jay Feldman, executive director of the Washington, D.C.-based group Beyond Pesticides/National Coalition Against the Misuse of Pesticides.

A 2000 square foot home requires that 380 gallons of pesticide be pumped into the ground. In a 100 home subdivision, that's about 38 thousand gallons put where children and pets play, and the family gardens.

E-Wire Press, "Hidden Pesticide Hazards Lurk
in Newly Built Homes"
17 Apr 02

Unborn U.S. babies are soaking in a stew of chemicals, including mercury, gasoline byproducts and pesticides, according to a report to be released Thursday. The report by the Environmental Working Group is based on tests of 10 samples of umbilical cord blood taken by the American Red Cross. They found an average of 287 contaminants in the blood, including mercury, fire retardants, pesticides and the Teflon chemical PFOA. "These 10 newborn babies were born polluted," said New York Rep. Louise Slaughter, who planned to publicize the findings at a news conference Thursday. "If ever we had proof that our nation's pollution laws aren't working, it's reading the list of industrial chemicals in the bodies of babies who have not yet lived outside the womb."

Maggie Fox, "Unborn Babies Soaked
in Chemicals, Survey Finds"
Reuters 14 Jul 05

The nationwide price tag of perchlorate cleanup could be in the tens of millions, and possibly even billions, of dollars, according to water officials and other experts, who say it has the potential to dwarf California's problems with MTBE, a gasoline additive that tainted groundwater supplies. Perchlorate, which is highly soluble, has been detected in water supplies in California and at least 19 other states, usually near defense contractors or military bases. The Colorado River, which supplies drinking water to about 15 million people in the

Southwestern United States, contains perchlorate that leached from the site of a former Nevada rocket fuel factory.

Miguel Bustillo, *"Lettuce Samples Found Tainted"*
Los Angeles Times 28 Apr 03

According to the latest data available from the American Wood Preservatives Institute's 1995 statistical report, about 1.6 billion pounds of wood preservatives are used to treat wood each year, including 138 million pounds of CCA, 656 million pounds of penta, and 825 million pounds of creosote. The three wood preservatives targeted by the lawsuit are linked to a wide range of health problems including cancer, birth defects, kidney and liver damage, disruption of the endocrine system and death. Two of the components of CCA, arsenic and chromium (VI), are classified as known human carcinogens. Creosote, a mix of toxic chemicals, is a cancer causing agent and it can cause nervous system damage.

Cat Lazaroff, *"U.S., Canada Groups Sue Over Toxic Wood Preservers"* *Environment News Service*, 11 Dec 02

Birds are being affected by lead on a massive scale. As of February 4, more than 176 trumpeter swans have been picked up dead or dying on the ponds they use in northern Washington State…it takes only three or four lead pellets to cause lead poisoning in a swan. Lead shotgun shells used for hunting contain about 280 lead pellets. For years, duck hunters left about 6,000 tons of lead shot annually in United States ponds, lakes and rivers before the US Fish and Wildlife Service banned its use in waterfowl hunting.

Jackie Alan Giuliano, Ph.D., *"Missing the Target - Green Bullets"*
Environment News Service, 27 Jun 01

More than a quarter of the world's coral reefs have been destroyed by pollution and global warming, experts said Monday, warning that unless urgent measures are taken, most of the remaining reefs could be dead in 20 years. In some of the worst hit areas, such as the Maldives and Seychelles islands in the Indian Ocean, up to 90 percent of coral reefs have been killed over the past two years due to rises in water temperature. Coral reefs play a crucial role as an anchor for most marine ecosystems, and their loss would place thousands of species of fish and other marine life at risk of extinction.

Associated Press in the Deseret News,
"Coral reefs in grave peril, scientists say"
24 Oct 00

Frogs exposed to a mix of pesticides at extremely low concentrations like those widely found around farms suffer deadly infections, suggesting that the chemicals could be a major culprit in the global disappearance of amphibians, UC Berkeley scientists reported Tuesday. When tadpoles were exposed in laboratory experiments to each pesticide individually, 4 percent died before they turned into frogs. But when atrazine and eight other pesticides were mixed to replicate a Nebraska cornfield, 35 percent died. At least one-third of amphibians worldwide, or 1,856 of the known species of frogs, toads, salamanders and caecilians, are in danger of extinction, according to an international group of conservation biologists.

Marla Cone, *"A New Alarm Sounds for Amphibians"*
Los Angeles Times, 25 Jan 06

Most oil pollution in North American coastal waters comes not from leaking tankers or oil rigs, but rather from countless oil-streaked streets, sputtering lawn mowers and other dispersed sources on land, and so will be hard to prevent, a panel convened by the National Academy of Sciences says in a new report. The thousands of tiny releases, carried by streams and storm drains to the sea, are estimated to equal an Exxon Valdez spill — 10.9 million gallons of petroleum — every eight months. When fuel use on water, either inland or offshore, is also taken into account, the report says, about 85 percent of the 29 million gallons of marine oil pollution in North America each year comes from users — drivers, businesses, boaters — and not from the oil industry. In particular, spills from tankers, barges and other oil transport vessels totaled less than a quarter-million gallons in 1999, down from more than six million in 1990.

Andrew C. Revkin, *"Offshore oil pollution*
comes mostly as runoff"
New York Times, 24 May 02

187

The hole in the ozone layer over Antarctica is now three times larger than the United States — the biggest it's ever been, scientists at NASA said Friday. In a sign that ozone-depleting gases churned out years ago are just now taking their greatest toll, this year's South Pole ozone hole spreads over about 11 million square miles. Too much UV radiation can cause skin cancer and destroy tiny plants at the beginning of the food chain.

Associated Press reported on MSNBC.com,
"Largest ozone hole on record spotted" 8 Sep 00

Thousands of idled farm workers [many from Mexico] are facing hunger in what has been a major center of California agriculture, hit by drought and court-ordered cutback of water supply to agriculture to protect endangered species (such as the Delta smelt). Water shortages cost California agriculture $260 million in 2007 as water deliveries fell to 35 percent of the routine level. The water shortage is expected to erase more than 55,000 jobs in the San Joaquin Valley. California has exceeded its sustainable water usage level for years, as its population rose from 10 million to 36 million in the last century.

Tracie Cone, *"Food Scarce Even in US Produce Capital; Water Shortage Worst in 3 Decades"* Seattle Post-Intelligencer, December 13, 2008. p. A4.

Industrial vomit fills our skies and seas while pesticides and herbicides filter into our foods. Twisted automobile carcasses, aluminum cans, non-returnable glass bottles and synthetic plastics form immense middens in our midst as more and more of our detritus resists decay. We do not even begin to know what to do with our radioactive wastes—whether to pump them into the earth, shoot them into outer space, or pour them into the oceans. Our technological powers increase, but the side effects and potential hazards also escalate.

Alvin Toffler, *Future Shock*, 1970

SECTION VII: ACTIONS AND SOLUTIONS

CHAPTER 39: INDIVIDUAL SOLUTIONS TO OUR DILEMMA

"If you want something done right, you have to do it yourself."

My Dad

An obvious question emerges: what fundamental component of our overpopulation crisis must we solve before our society and the rest of the world stands any chance of surviving this global "Human Katrina" population disaster?

Primary answer—in addition to sensible immigration control actions: *an alternative energy source to oil, coal and natural gas. An alternative energy source holds the key to survival for advanced civilizations on this planet.*

In order to create collective action to change current immigration inflow, the following organizations stand on the front line. Join them for free and take actions via their alerts. If you do nothing else, join www.NumbersUSA.com and add yourself to nearly one million members. You will be directed to send pre-written electronic faxes along with phone calls to bring immigration back to sustainable numbers. Tell all your friends!

1. Join www.NumbersUSA.com with Roy Beck as one of the most effective and successful non-partisan organizations. Become a member of his fax/phone call armada of active citizens. Membership exceeds one million people, which harnesses a great deal of collective power. Distribute his video *"Immigration by the Numbers"* on the web site to educate your

networks as to the dangers of mass immigration. I highly recommend joining this organization.

2. Join www.thesocialcontract.com with Dr. John Tanton in order to educate yourself as to the consequences of overpopulation. Additionally, subscribe to *"The Social Contract Quarterly"* that provides you with the latest information and scientific update on population dynamics. Distribute the booklet: *"Common Sense on Mass Immigration."* I highly recommend joining this organization.

3. Join www.fairus.org with Dan Stein at Federation of American Immigration Reform. He promotes sensible immigration reform that curtails mass immigration. My highest recommendation.

4. Join www.capsweb.org with Dr. Diana Hull in California. Californians for Population Stabilization stands at the forefront in our most overpopulated state. Dr. Hull offers faxes and phone calls that empower citizens everywhere in the United States to take action on a proactive level. Very powerful and empowering to average citizens! My highest recommendation for joining this organization.

5. Join www.alipac.us with William Gheen to keep you on the front line of this great national struggle. His member updates encourage local activism with the tools to make impact.

6. Join www.firecoalition.com with Jason Mrochek to bring information and tools to your fingertips. He stands on the front lines in California with members all over the country. Within that web site, you will find a half dozen other effective web sites to take action.

7. Join www.grassfire.org with Steve Elliot for up-to-date information and actions you can take in your local areas.

8. Join www.cis.org or Center for Immigration Studies with Mark Krikorian for timely information on U.S. immigration impacts. Read Krikorian's *The New Case Against Immigration: Both Legal and Illegal.*

9. Join www.limitstogrowth.org ; www.immigrationshumancosts.org with Brenda Walker to bring

you outstanding information and data mostly squelched by main stream media.

10. Join www.worldpopulationbalance.org with David Paxson and his team to keep you apprised of the global impacts of population.

11. Join www.populationmedia.org with Bill Ryerson for a firsthand look at what other countries experience as to overpopulation.

12. Join www.vdare.com with Peter Brimelow for outstanding essays and information on what we face as a civilization if we allow continued massive and unending immigration.

13. Join www.carryingcapacity.org and www.balance.org with David Durham for up to date and outstanding information and action items that you may use for very effective impact. Durham has worked for 25 years on this issue and works to bring about sensible immigration to sustainable levels.

14. Join www.transitionus.ning.com ; www.transitiontowns.com to prepare yourself and your community for the post Peak Oil descent.

15. You will find excellent information in the following: www. plannedparenthood.org ; www.populationconnection. org ; www.care2.com ; www.world.org ; www.npg.org ; www.populationaction.org ; www.projectusa.org ; www. populationconnection.org ; www.proenglish.org ; www. optimumpopulation.org

16. Important websites:

    Global Public Media; www.globalpublicmedia.com
    Energy Bulletin; www.Energybulletin.net
    The Oil Drum; www.theoildrum.com
    Post Carbon; www.postcarbon.org

Most of the aforementioned organizations offer speakers to invite to your local area.

The following information may be found at www.frostywooldridge.com for this action letter that you can forward to your network of friends. They, in turn, can enlarge the network via their friends. This action letter covers individual and community actions available to all citizens.

ACTIONS YOU CAN TAKE TO CHANGE OUR COUNTRY TOWARD A SUSTAINABLE FUTURE!

Educate yourself on what we face and take action by joining these non-partisan organizations that empower you personally and collectively: www.numbersusa.com ; www.thesocialcontract.com ; www.fairus.org ; www.capsweb.org ; www.firecoalition.com; www.alipac.us ; www.cis.org

Contact top national shows by suggesting they interview experts on overpopulation such as Lester Brown, Fred Meyerson, Dr. Diana Hull, Dr. Albert Bartlett, Governor Richard D. Lamm, Lindsey Grant, Kathleene Parker, Don Collins, Dan Stein, William Gheen, David Paxson, Buck Young, Frosty Wooldridge, Bill Ryerson, Jason Mrochek, and many others.

Write to: charlierose@pbs.org ; oreilly@foxnews.com ; lou.dobbs@mail.cnn.com ; morning@npr.org ; 60m@cbsnews.com ; hannity@foxnews.com ; Katie Couric at evening@cbsnews.com ; earlyshow@cbs.com ; special@foxnews.com ; today@nbc.com ; dateline@nbc.com ; letters@newsweek.com ; letters@time.com ; letters@usnews.com ; editor@usatoday.com ; www.nightly@nbc.com ; late.edition@cnn.com ; Bob Schieffer at ftn@cbsnews.com; David Gregory at MTP@nbc.com ; Chris Mathews at cmatthews@msnbc.com

Environmentally, we must take action as to water, energy, air and land: www.wateruseitwisely.com ; www.treehugger.com ; www.willyoujoinus.com www.populationmedia.org ; www.ases.org for energy; www.greenpeace.org ; www.stopglobalwarming.org ; www.greenprintdenver.org ; www.mcs-global.org ; www.savetheoceans.org ; www.optimumpopulation.org

What we must discuss: "U.S. Sustainable Population Policy"; "U.S. Carrying Capacity Policy"; "U.S. Environmental Impact Policy"; "U.S. Water Usage Policy." We must shift toward a sustainable, stable population for our civilization.

Personal actions: letters to the editor, speak at city council meetings to advance this information to fellow citizens, call local TV stations and ask to have state leaders on this issue interviewed in your state.

Arrange for a 15-20 minute meeting with your U.S. House Rep, U.S. Senator, State House Rep and State Senator. Inquire at my website www.frostywooldridge.com for an outline covering that 20 minute presentation. Call into local radio stations in your community to take up the issue by interviewing experts on limits to water, clean air and highways.

Form clubs, groups and web sites that make impact at the local level. Engage the media to report in your local papers.

Encourage other groups to host Frosty Wooldridge's "*THE COMING POPULATION CRISIS IN AMERICA: WHAT YOU CAN DO ABOUT IT*" or Roy Beck's "*IMMIGRATION BY THE NUMBERS*" or David Paxson "*World Population Crisis.*" www.frostywooldridge.com ; www.numbersusa.com ; www.worldpopulationbalance.org. Go to these websites for more action items.

Subscribe to E-Magazine, OnEarth, Greenpeace Magazine and/ or other environmental magazines that will inform you. Type into Google-- "The top 100 environmental websites." You will receive information on the top 100 most effective websites on how you can take action.

These above actions will change consciousness in this country from unlimited growth and population gains to a stable, sustainable population. We must create a 'critical mass' of Americans that push this issue to the top of the presidential and congressional agendas. By gaining millions, we reach 'tipping point' that changes history.

Who is your hero? Susan B. Anthony? Barack Obama? LeBron James? Dr. Martin Luther King? Gandhi? Eleanor Roosevelt? Michael Jordan? John Muir? Teddy Roosevelt? Cesar Chavez? Barbara Jordan? Federico Pena? They were common citizens with uncommon determination. That was their time, this is yours. You possess the same power to change history for the better. You are invited to bring your relentless enthusiasm and passion for your country, for your planet and in the end—for your children—to this noble task before all humanity.

Join thousands of citizens by becoming a national citizen press corps agent—with guidelines to engage media to interview leaders on overpopulation. For details, contact: frostyw@juno.com ; www.frostywooldridge.com

Books to inform:

Too Many People by Lindsey Grant
A Bicentennial Malthusian Essay by John Rohe
The End of Nature by Bill McKibben
Out of Gas: The end of the Age of Oil by David Goodstein
The Long Emergency by James Howard Kunstler
The Population Bomb by Paul and Anne Ehrlich
Stalking the Wild Taboo by Garrett Hardin
The New Case Against Immigration: Both Legal and Illegal by Mark Krikorian
Our Plundered Planet by Fairfield Osborn
The Sixth Extinction by Leakey and Lewin
Food, Energy and Society by David and Marcia Pimentel
Biggest Lie Ever Believed by Michael Folkerth
Shoveling Fuel into a Runaway Train by Dr. Brian Czech
Peak Everything: Waking up to the Century of Declines by Richard Heinberg
The Population Fix: Breaking America's Addiction to Population Growth by Edward C. Hartman
The Transition Handbook: From Oil Dependency to Local Resilience by Rob Hopkins
Overloading Australia by Mark O'Connor and William J. Lines

Bio-Centric Imperative by William Dickinson
DVD: *Blind Spot www.blindspotdoc.com*

What can you do to bring change?

You must become part of critical mass—enough citizens writing, calling, talking and promoting anyone who addresses this issue at the local, state and national levels.

Our leaders have lost touch with you, and reality. They need your help.

We must spend billions for research on alternative energy to free us from the grips of petroleum-oil. The thinking that got us into this mess cannot be used to get us out. New paradigm brain-storming must begin.

We cannot keep thinking that the solutions of the 20th century will solve our problems in the 21st century. They will not.

We must find solutions that work for the future of all humanity

We must stop fossil fuel burning as soon as possible. We must investigate both sides (in balance) of the climate change question, as soon as possible, then take decisive action and get to work on real solutions.

We must stop massive species extinction as soon as possible

We must balance our population— inside the carrying capacity of our nation and the planet— in order to bring about a sustainable society moving into the future.

Please, add your ideas to this basic list:

1. Possible alternatives show promise in solar conversion. Let's shift gears and engage ten times more top scientists to explore sunshine as our best non-polluting source of energy by changing it from light beams to electrical power. Let's engage wind power and wave power toward electrical power.

Visit American Solar Energy Society in Boulder, Colorado at: www.ases.org.

2. Learn about the National Energy Research Laboratory (NREL) research and development of renewable fuels and electricity that advance national energy goals to change the way we power our homes, businesses, and cars.

   NREL's brand new "Wind to Hydrogen" facility offers a new template for future energy production. Xcel Energy and NREL recently unveiled a unique facility that uses electricity from wind turbines to produce and store pure hydrogen, offering what may become an important new template for future energy production.

3. Produce only two and four cylinder cars. Nobody needs a six or eight cylinder car. I've been driving four cylinder cars for a long time— starting with my VW in 1965. I've always arrived at my destination safely, and on time.

4. To promote more efficiency in the near future, we might promote possible incremental incentives:
   * Charge $8.00 per gallon for all eight cylinder cars
   * $6.00 a gallon for six cylinder cars
   * $2.00 a gallon for four cylinder cars
   * $1.00 a gallon for two cylinder vehicles
   * Press for more hybrid cars, with 70-100 miles per gallon.

5. From that pricing-incentive approach, everyone would start moving toward greater efficiency, bicycling, car pools, bus, mass transit and ecological responsibility. If you think these ideas sound harsh, just wait until the oil runs out— for greater trauma. If you've got better ideas, let's hear them! We must stop green house gas emissions: www.stopglobalwarming. org. Please, stop and remember—more people make more greenhouse gases.

6. Create a nationwide bottle, can and plastic 10 cent deposit/ return on every container sold at retail. That includes every container sold out of every store in America! No one loses a dime, and everyone becomes skilled recyclers! Money speaks louder than words.

7. Waste disposal: average Americans create around five pounds of waste daily. You may choose a number of waste reducing methods.
    a. Bring your own cotton bags to grocery and retail outlets. I have cotton bags 27 years old and still hauling groceries!
    b. Vote to stop junk mail. www.stopjunk.com
    c. Use rechargeable batteries.
    d. Buy "green" with less packaging. For more information, visit www.greenprintdenver.org for new ideas and how to start in your community. Use that web site for a template in your own state. Also, www.treehugger.com is an excellent site for environmental ideas and solutions.
8. How do you stop using chemicals that harm you and your family?
    a. Avoid any household product marked with "Caution"; "Warning"; "Danger".
    b. Use earth-friendly cleaners, purchased at your local stores.
    c. Dispose of hazardous waste by taking to eco-recycle experts.
    d. Remember that everything you store in the house outgases to you and your children. Keep such chemicals in your garage.
    e. www.ecos.com; www.melaleuca.com; www.7thgeneration. com for safe cleaning products for your home. Also, www. mcs-global.org helps you deal with chemical poisoning.
9. We face a dreadful water crisis if we fail to stabilize our population. If we stopped all immigration today, our population momentum will add more than 35 million within 40 years. Therefore, we must deal with our water issues.
    a. Practice water conservation with low flow flush toilets, water saver shower heads and know that every time you leave the tap running, it wastes water. www. wateruseitwisely.com
    b. Create xeriscape lawns-landscaping with minimum water needs. www.greenprintdenver.org and www.treehugger. com

    c.   Move toward a healthier nutritional program by shifting toward more fruits, vegetables, grains and alternative protein products such as rice, beans, tofu and soy products to avoid more than 1,000 gallons of water and 16 pounds of grain used to produce one pound of beef.

10. Enjoy this revitalized thought-process; plug into like-minded people determined to make this planet livable. www.emagazine.com will give you up to the minute information. Visit: www.thesocialcontract.com and become a member and enjoy the Social Contract Quarterly for incredible information on conferences, writers and books.

11. Introduce environmental plans and ideas at your city council, church group and civic clubs.

12. Invest in solar power, wind power and other alternatives. Invest in solar heating panels on the roof of your house for heating water and rooms.

13. If we fail to find another energy source to power our infrastructure—to run our cars, power plants and home heating—energy dependent civilizations around the planet will not survive. www.savetheoceans.org and www.greenpeace.org to promote health for our oceans, reefs and rainforests.

14. Visit www.recyclebank.com for information leading to recycling programs in your area or find out how to start them in your community.

Everything you do—counts!

Visit www.world.org for 100 top web sites dealing with every method you can imagine for making your world a better place in which to live.

To each passionate citizen reading this final part of the book, I invite you to act. If you take little or no action, your inaction creates impact— think about it!

Live in a state with lots of rivers? Protect them by adding your voice to www.americanrivers.org

Finally, 900 cities around the world promote a new movement for all humanity. They move toward self-sustaining communities based on local production of food, shelter and goods. You may work toward a sustainable community in your area in America by contacting <u>www.transitionus.ning.com</u> . As our oil reserves decline, everyone will be forced to move toward smaller local communities that remain self-sustainable. The aforementioned website offers a complete protocol for nearly every state in the U.S.

Each day I sit in front of my keyboard, I read aloud the following to maintain momentum. I share it with you as we come together in this urgent action:

"Until one is committed, there is hesitancy,
the chance to draw back—always ineffectiveness.
Concerning all acts of initiative and creation, there is one elementary truth—
the ignorance of which kills countless ideas and splendid plans.
That the moment one definitely commits oneself to a task,
then Providence moves too—acting as one.
All sorts of things occur to help—that would never otherwise have occurred.
A whole stream of events issues from the decision, raising in one's favor;
All manner of unforeseen incidents and meetings and material assistance,
which no one could have dreamed would have come her/his way.
Whatever you can do, or dream you can do, begin it.
Boldness has genius, power and magic in it.
Begin it now!"

Goethe (German philosopher and Murray)

All living creatures on earth thank you for your actions!

If you would like to book this program in your city, please contact as shown below:

Title of the presentation: "COMING POPULATION CRISIS IN AMERICA: WHAT YOU CAN DO ABOUT IT"

The USA will double its population from 300 million to over 600 million sometime past mid century—if current growth rates continue. It will add 100 million within 32 years by 2040. Demographic predictions show Colorado adds five million people to the Front Range in the next 40 years. California will add 20 million within the next 30 years. Texas will add 12 million by 2025. We create an irreversible crisis with unsolvable problems for our children.

"The problems in the world today are so enormous they cannot be solved with the level of thinking that created them." Einstein

Frosty Wooldridge, former Colorado math/science teacher, has bicycled 100,000 miles on six continents and six times across the United States coast to coast in the past 35 years. He has witnessed the crisis of overpopulation in Mexico, China, Bangladesh, India and South America.

He presents a clear picture of the future for America if we continue on this Titanic-like course while offering solutions that stand in line with Einstein's appreciation for stepping out of the box. Wooldridge began as a teacher in Brighton, Colorado since 1973 and is the author of seven books. He presents a powerful and compelling program to all audiences laced with humor, compassion and sense of optimism. Wooldridge has been a guest on hundreds of radio and TV programs across the United States including ABC, CBS, NBC and FOX.

As Eleanor Roosevelt said, "We must prevent human tragedy rather than run around trying to save ourselves after an event has already occurred. Unfortunately, history clearly shows that we arrive at catastrophe by failing to meet the situation, by failing to act when we should have acted. The opportunity passes us by and the next disaster is always more difficult and compounded than the last one."

Excellent references by writing former
Colorado Governor Richard D. Lamm

Ken Hampshire, President of Advanced Health
Group <u>kenh@advancedhealthgroup.com</u>

Donald McKee, physicist <u>mckeedon@gmail.com</u>

Choose the 25 minute program or the 45 minute
program that includes dramatic world travel slide show
that takes audiences from the Arctic to Antarctica.

Wooldridge distributes information
whereby every person in the audience may
take action on a personal level to ensure a viable
future for our children and the United States.

Frosty Wooldridge
POB 207
Louisville, CO 80027
<u>www.frostywooldridge.com</u>

CHAPTER 40: ACTIONS AT THE NATIONAL LEVEL

"In the current instance, the rational basis for the appeal, and its centrality to our survival, are clear. Nothing is to be lost and everything to be gained by sharing accurate and relevant information about our situation; there is no need to exaggerate the threat."

Richard Heinberg, Peak Everything

Where do we start? What on earth can anyone do to change the grave results of adding 100 million people to the United States by 2035? What about the fact that the rest of humanity expects another three billion added by 2050? Why won't our leaders speak up about an issue so dire that it affects every American today and in the future?

Why wouldn't we question U.S. population growth when millions sit in gridlock traffic, breathe toxic air, pay more for everything, suffer water shortages—and know that it can only worsen?

Heck of a question! Everyone runs from the answer

As you can imagine—religions, emotions, cultures and history lock most humans into paradigms formed 2,000 years ago. While such paradigms fail in the 21st century, most humans cling to those beliefs as a life raft in the desert when they must search for water. Capitalism, corporations and religions' premises that humans can multiply forever shall prove our greatest obstacle to reasonable choices.

As you read this information, you realize our civilization stands at risk. What can you do?

In all of recorded history, passionate men and women rose out of nowhere to take action to right some wrong. Some became famous and most did not. Fame meant nothing to them other than it became a byproduct of their actions. Another aspect of their work rendered what Malcolm Gladwell called a "tipping point" in history where change occurred via the bravery and passion of common persons with uncommon determination.

That "tipping point" became critical mass that provoked American colonists to fight and die for their new country. Susan B. Anthony provided a "tipping point" to gain voting rights for women. In India, Gandhi's walk to the sea provided a "tipping point" to oust the British. Dr. Martin Luther King provoked the civil rights movement.

Which great person in history inspires you? John Muir? Betsy Ross? Teddy Roosevelt? Charles Lindbergh? ML King? Barack Obama? Michael Jordan? Amelia Earhart? Michelle Obama? Harry Truman? Your dad? Your mom? What do they all possess in common? Noble purpose!

That was their time; this is yours

I invite you to bring your highest creative energy to this noblest hour in history. What we do in the next decade will change the course of history for America and the world. That can mean for better or worse, depending on our collective actions.

How can we stop the United States from adding 100 million people?

In this chapter of the book, we'll examine logical, practical and reasonable choices to stabilize America's population to a sustainable future for all citizens. By taking actions for ourselves, we will inspire and invite other countries to establish their own choices.

Since the U.S. female enjoys a 2.03 fertility level, it's not America growing its population from within. What causes our accelerating population growth? Short answer: both legal and illegal immigration.

Since you own or rent your dwelling, you may choose to invite 20 guests or 100 guests depending on your available space. We as a nation must choose how many guests we can sustain in our "house."

Since legal and illegal immigration constitute the main sources of our population crisis, we must act to diminish the cause of over 2.4 million immigrants added annually.

Yes, our actions will force overloaded nations to reconsider, change and solve their own population challenges. Is it unreasonable for national and personal responsibility?

We must consider and engage:

1. A TEN YEAR MORATORIUM ON ALL IMMIGRATION: This would allow our country to regain its collective breath. It would allow us to regain our schools, language, medical facilities, financial balance, ecological viability and order, which is necessary for a first world country to operate for all its citizens.

We must employ a linkage strategy. In other words, we must create a paradigm shift that employs all the following actions to reap a plausible future for humans in America. A former congressman said, "The challenge is enormous and you have to talk about a moratorium. You can't talk about anything short of a moratorium because, frankly, anything less will never get you one step closer to population stabilization."

In addition to a moratorium, we must engage a "Sustainable Immigration Policy" with a maximum of 100,000 immigrants annually—with needed skills to our benefit—that speak the English language before they arrive—in order to be considered for entry into the United States. If that maintains our stable population, we can continue. If not, we must maintain a moratorium.

Why the number 100,000 allowed into the USA annually? That's how many people egress the United States each year. (Higher or lower depending on the year.) Thus, we would enjoy a net gain of zero! Whatever our annual immigration numbers into this country in the years ahead, we must balance it with humans leaving our country in order to maintain a stable and sustainable population.

We could entertain a farm guest worker program only if it stipulates that male workers enjoy an entry date for three months to a maximum of six months and an exit date. No female or family members allowed. Additionally, no 'anchor babies' or 'instant citizenship' allowed for foreign nationals' babies born on U.S. soil.

Yes, we must work with Americans marrying foreign spouses and a few other visa considerations, but we must hold to our limited carrying capacity.

2. A NATIONAL SUSTAINABLE POPULATION POLICY: We must create a strategic plan or vision for the future of the United States of America. How many people can this nation hold and still maintain our standard of living and quality of life? How can we maintain it in order to provide the "American Dream" opportunity for all citizens? How can we maintain enough water and farmland to feed ourselves instead of depending on elusive imports from other countries as climate change and gasoline become more critical? What is our population limit?

Some people will make the charge, "That's population control! How terrible!" Thankfully, since 1970, the U.S. female has averaged 2.03 children, thus a steady and balanced population.

Additionally, other western countries enjoy replacement birth rates, but suffer overwhelming immigration rates that exceed their carrying capacities. The United Kingdom expects an additional 11 million people within two decades, virtually all of it, via immigration. They cannot possibly survive such an onslaught. Already over their carrying capacity at 21 million people on a continent that features 96 percent desert, neither can Australia! Therefore, you may engage with the following organizations.

Websites: www.nationaloptimumpopulationcommission.com (US)
www.optimumpopulation.org (United Kingdom)
www.population.org.au (Australia)
www.populationinstituteofcanada.ca (Canada)

For Canadian readers, the USA and Canada share a common history and common fate. Tim Murray leads Canada's movement against mass immigration causing their hyper-population growth.

> Tim Murray, Director Immigration Watch Canada
> www.immigrationwatchcanada.org
> Vice President, Biodiversity First
> www.populationinstituteofcanada.ca
> www.immigrationwatchcanada.org
> http://biodiversityfirst.googlepages.com/index.htm
> http://sinkinglifeboat.blogspot.com
> http://ecologicalcrash.blogspot.com
> http://sustainablesalmonarm.ning.com
> http://www.ramacresearch.ca/
> http://canadianimmigrationreform.blogspot.com/
> http://www.ramacresearch.ca/
> http://canadianimmigrationreform.blogspot.com/

We must encourage and maintain a two children or less family policy. We can do it by choice today because we already stand at 2.03 children per American female. We can utilize positive income tax deductions for two or less children and negative income tax for families that choose more than two children. If we wait, we'll be in the same boat as the Chinese with forced one child per family.

Of course, religious leaders will scream 'abortion' and other critical name calling. No! A population policy gifts us with choices for birth control before our civilization runs out of options. As it stands today, worldwide, women undergo 43 million abortions annually. A simple birth control program would go a long way in correcting that human tragedy. Or, do we and church leaders prefer millions dying of starvation annually as they wallow in misery from hyper-population growth?

In addition, we must enact a state by state policy—I live in Colorado, so a "Colorado Sustainable Population Policy" to determine how many people can live in Colorado within the water, food, land and resources in my state. Obviously California can hold more than Rhode Island, etc.

3. A NATIONAL ENVIRONMENTAL IMPACT POLICY: The U.S. and all countries must deal with the sobering reality of stabilizing human population. We must stop causing exploding extinction rates to all other creatures on this planet by our own expanding numbers. As head of the free world, we must lead in the understanding that an environmental impact policy will give other creatures and future generations a sustainable planet. Without it, we drive forward with no idea of where we're heading and where we'll end up. China already suffers horrible consequences and India is even worse because it keeps driving its civilization over a cliff as it adds another 400 million by 2050. We must not follow their paths!

Each state must implement an "Environmental Impact Policy" to determine how many people will allow enough land and habitat for animals in that state. Every state must develop these policies along with a "Water Usage Policy."

4. MANDATORY RECYCLING LAWS: Additionally, we must move toward a national 10 cent deposit/return recycling law on every piece of plastic, glass and metal sold out of retail stores to guarantee 99 percent recycling to stop the carnage of our environment with throwaways.

5. NATIONAL WATER CONSERVATION POLICY; NATIONAL ENERGY CONSERVATION POLICY; NATIONAL LAND CONSERVATION POLICY: We need to implement conservation policies that will directly affect future generations on a positive scale. To continue helter-skelter growth and expansion without a plan will create irreversible consequences and unsolvable problems for future citizens of the United States.

6. FAMILY PLANNING WORLDWIDE: The United States and other first world nations need to assist other countries

with family planning methods. Birth control is a major aspect of family planning. Without it, third world countries suffer endless population increases and degradation. During one of my lectures a student asked, "How can we force other countries to have two child families…isn't that overstepping our bounds?"

To answer that question, I said, "No, we can never infringe on other cultures or countries. We can only offer them the means and ability to bring their societies into population stasis. If they choose otherwise, they will find that nature proves to be the ultimate population arbiter."

Lester Brown, *Plan B 3.0,* wrote: "Population growth, which contributes to all the problems discussed here, has its own tipping point. Scores of countries have developed enough economically to sharply reduce mortality but not yet enough to reduce fertility. As a result, they are caught in the demographic trap—a situation where rapid population growth begets poverty and poverty begets rapid population growth. In this situation, countries eventually tip one way or the other. They either break out of the cycle or they break down."

To subsist, the Catholic Church, Protestants, Buddhists, Hindu, Islam and other religions that continually work against family planning must step out of the Dark Ages and into the 21st century. We can promote education that spotlights and unhinges their entrenched thinking based on concepts that formed 2,000 years ago. Humans can no longer usurp nature's laws via hyper-population growth.

It's not logical to think any of these great religions would transform toward rational action by accepting family planning any time soon. Therefore, we must take care of our citizens first to ensure our country's viability. After that, we may continue our assistance worldwide.

7. DEVELOPMENTAL ASSISTANCE: First world countries need to assist with development, housing, education, fresh water, family planning and health care for countries that suffer from this planet-wide population crisis. Help them in their own countries. That means tractors, crop techniques, irrigation, etc. No financial aid because money tends to go into the hands of the leaders.

8. ELECT SENATORS AND CONGRESSIONAL REPRESENTATIVES WHO REPRESENT AMERICANS: One of the reasons Congress and our presidents have created this national atrocity stems from their entrenchment in the 'good old boy' network. As long as they represent massive immigration as well as corporations who pander to this predicament, you will not see change. You must elect leaders who will take action on behalf of Americans, and really, humanity.

A scant 50 percent of Americans vote in national elections. Local elections rate less than 20 percent participation most of the time. More Americans see this crisis and step up to the plate. They must run for office in many states.

9.The most important factor for saving America, and really, the planet: you! Use the Internet with web sites that create collective action. Connect with all Americans! Use your money and your time! Stand up! Write! Call! Educate on radio talk shows! Call on TV networks and express your concerns. Express your ideas! Be heard! Be seen! Be passionate! Demand! Expect action! Become action in motion! Instead of watching the news, become the news by your actions!

CHAPTER 41: INTERNATIONAL ACTIONS

"All things are possible when enough human beings realize that everything is at stake."

Gary Snyder

This blue-green orb whirling through the universe proves an amazing creative wonder to any human being that looks up into the night sky. As the morning sunlight bursts across the eastern horizon, seldom do other species take special notice. But for humans, the drama unfolds in light banners, streamers, blazing clouds and an array of creative processes.

Perhaps our planet-home may be a fluke of the universe but, as much as we know, this sphere proves to be the only game in town.

Additionally, we find, as humans, our destructive actions threaten not only fellow life forms on this globe, but our long term existence.

In the past 100 years, we wreaked havoc on this planet with chemicals, fossil fuel burning and too many of our species. We conquered nature by destroying its balancing systems. As a consequence, today, our "Human Katrina" created a "Human Dilemma" from which we must engage "Human Solutions."

Otherwise, the nature of Mother Nature dictates harsher and harsher results.

What responsibility do we possess for future generations?

In a simple statement: we owe our progeny, as well as all living creatures on this spaceship-home—a livable, viable and sustainable planet that we inherited.

How can we do that as a collective effort worldwide?

1. Leaders from all countries or as many major nations as can be incorporated-- must form an international coalition of governments coming together with a theme of *"World Population Stability for a Sustainable Future for all Living Creatures on Planet Earth."*

2. They must create and evolve a viable plan for stabilizing world population via education, birth control and family planning. They must hammer out a consensus for future generations. They must gather the world's brightest minds to create an alternative energy source to supplant oil as soon as possible. They must address species extinction and climate change. They must move humanity toward quality of life and environmental balance while working with human and animal dignity.

3. The richest and/or most populated nations may host the conference(s) in order to create a world plan. If no nation will step forward to host such a conference, the United States must take the lead by inviting as many nations to this world summit as will accept.

4. While they may introduce several viable paradigms, Lester Brown, president of Earth Policy Institute, wrote, *PLAN B 3.0,* which offers one of the finest, most refined approaches for saving civilizations around the planet.

5. Once a 'critical mass' of nations sign on with, *PLAN B 3.0,* or some form thereof, humanity can move toward a sustainable future based on population stability and 'steady state economics'.

Think back 25 years! What did you do with your life? What meaningful triumphs? What childhood bliss did you enjoy? How about the high points of your high school or college years or first job? Later, picture your kids graduating from high school or college. How about that trip to your favorite destination? How quickly did 25 years slip into your scrapbooks?

How old will you be 25 years from now? Forty-five? Fifty-five? Sixty-five? How much space, resources and energy that paint your 'ideal' future will have been constricted by hyper-population growth—if you and your government fail to act?

Whether we like it or not, our civilization depends on choices manifested by every single citizen's individual actions today. I trust this book enlightened you and catapulted you to take action at whatever level you feel empowered—so your dreams or your children's dreams 25 years from now may enjoy scrapbook memories.

NOTES

Foreword—

1. Mexico's growth rate 100 million to 153 million—UN Population Estimates
2. Time Magazine: "*How to end poverty*"—Eight million adults die each year because they are too poor to stay alive. March 14, 2005; By Jeffrey D. Sachs
3. Ten million children deaths annually: World Health Organization
4. *Camp of the Saints* by Jean Raspail 1973

Chapter 1: Consequences of Human Katrina

1. USA to add 138 million by 2050: "*U.S. Hispanic population to triple by 2050*" USA Today; by Haya El Nassar; http://www.usatoday.com/news/nation/2008-02-11-population-study_N.htm February 12, 2008
2. California population 2008: 37.5 million www.capsweb.org : U.S. Census Bureau

Chapter 3: Bow of the Titanic

3. Californians for Population Stabilization: Director, Dr. Diana Hull.
4. Federation for American Immigration Reform: Fogel/Martin, March 2006, "US Population Projections"

Chapter 4: For Lack of Water

5. Newsweek Magazine; April 16, 2007: "Leadership and Environment"

Chapter 5: Losing the Wild

6. *Foundation of Economic Trends*: Jeremy Rifkin

Chapter 6: Crossing Our Agricultural Rubicon

7. The Social Contract; "*Crossing our agricultural Rubicon*" Social Contract Quarterly" spring 2005, Dr. John Tanton

Chapter 7: Science, Resistance and Human Constraint

8. Known as an 'ecological footprint' in the USA equals 12.6 acres of land used to support every American; www.socialcontract.com; www.cis.org ; Environmental Magazine

Chapter 10: Piling up on the Rocks

9. *Bio-Centric Imperative: How Population, Environment and Migration Shape Our Future* by William B. Dickinson

Chapter 12: Fracturing America

10. California population 2008: 37.5 million, growing by 1,700 per day. www.capsweb.org , Dr. Diana Hull, Director
11. "*French Muslim Riots Growing More Violent*" ; Outside the Beltway, James Joyner, November 5, 2005 http://www.outsidethebeltway.com/archives/french_muslim_riots_continue_spread/
12. Koran or Quran: Islamic 'Bible'; Sura (chapter) 9, verse 5—"Then fight and slay the Pagans (non-believers) wherever you find them. Seize them, beleaguer them and lie in wait for them, in every stratagem of war." Islam insists that all nations must be fought "until they embrace Islam."

Chapter 14: Energy and the Silent Lie

13. "*Lie of Silent Assertion*" by Mark Twain: Samuel Langhorne Clemens (November 30, 1835 – April 21, 1910)

14. Kenneth Ewart Boulding; "Dismal Theorem" (January 18, 1910–March 18, 1993) was an economist, educator, peace activist, poet, religious mystic, devoted Quaker, systems scientist, and interdisciplinary philosopher. He was cofounder of General Systems Theory and founder of numerous ongoing intellectual projects in economics and social science.

Chapter 15: Societal Self Destruction

15. Albert A. Bartlett; emeritus Professor of Physics at the University of Colorado at Boulder, USA. Professor Bartlett has delivered over 1,500 lectures on "*Arithmetic, Population, and Energy.*"

Chapter 16: How to Destroy America

16. Richard D. Lamm, Governor of Colorado 1975-1987, speech at F.A.I.R. conference, October 3, 2003, "*How to Destroy America.*"

Chapter 17: Death, Disease, Consequences

17. Rocky Mountain News; "*What happened?*" Denver's graduation gap—5,663 students started eighth grade in 1999; 1,884 graduated from Denver Public Schools in 2005, 67 percent dropout rate, May 16, 2005.
18. Time Magazine; "*America's Border: After 9/11, it's outrageously easy to sneak in*" Pulitzer prize winner Donald L. Barlett and James B. Steele, September 20, 2004
19. Source for TB in Michigan: www.clickondetroit.com, 11/4/03, "*Thumb Schools Report 34 Tuberculosis Infections.*"
20. Source: Mother Jones News, March issue 2003, Dr. Kevin Patterson, "*The Patient Predator.*"
21. Source for TB in Maine: www.news.maintoday.com 11/6/03, "*Health Officials Fight TB Outbreak*"
22. Source: Santa Barbara News-Press; "*Anatomy of an Outbreak*", by Melinda Burns, April 25, 2004

23. Source for Hepatitis: *"Shots for nearly 3,000"*: www.post-gazette.com, 11/6/03, *"Hepatitis Outbreak"*

24. Source for leprosy: Sharon Lerner, NY Times, 2/20/03, *"Leprosy on the rise in the U.S."*

25. Source: Carlos Bastien; *Chagas Disease: The Kiss of Death.*

26. Source: Hemorrhagic Fever, Dengue Fever, Head lice, www.fairus.org

27. Source: *Timebomb: Global Epidemic of Multi-drug Resistant Diseases* by Dr. Lee Reichmann

Chapter 18: Killing Our Oceans

28. Source: Paul Miller, Denver Post, *"The Plastic's Paradox"* January 13, 2007.

Chapter 22: Coming Mega-Traumas

29. Source: New York Times: *"As jobs vanish and prices rise, food stamp use nears record levels—28 million Americans"* by Erik Eckholm, March 31, 2008.

30. Dead Zones: http://daac.gsfc.nasa.gov/oceancolor/scifocus/oceanColor/dead_zones.shtml National Aeronautics and Space Administration, Mississippi River, September 20, 2008.

Chapter 23: Examples of our Future Abound

31. World population: Bill Ryerson, www.populationmedia.org

Chapter 25: Climate Change Starts

32. Professor John Schellnhuber; director of the Potsdam Institute for Climate Change Research, described Lovelock as "one of the most influential scientists on the environment for many years now" whose views have to be taken "very seriously."

Chapter 34: How and Why Journalists Ignore Population Connection

33. *The Population Explosion* by Paul and Anne Ehrlich, 1990, Touchstone Book with Simon & Schuster.

Chapter 36: Let's Change Direction on Human Population

34. *Stalking the Wild Taboo* by Dr. Garrett Hardin, 1996, The Social Contract Press

EPILOGUE

For certain, I'll be the first to admit, "Houston, we have a problem!" But this is not Apollo 13 with three crewmen facing imminent death out in the black void of space between Earth and the moon. We confront a dilemma with our entire planet and 6.7 billion humans along with all other life forms.

Everything described in this book will catch up to us in one way or another. Mine is not the only book exposing our gargantuan problems. I may, however, be one of the few to see those problems up close and ugly in world bicycle travels.

Within two decades, we most certainly will experience monstrous changes impacting us, our families and our communities.

Can humanity marshal its leaders quickly enough to decelerate hyper-population growth? Can world religions change their dogmas on birth control?

Can we discover an alternative energy source for whatever population accumulates? Can we save our planet-home from climate destabilization by the hands of humans?

It's going to be a horse race as to when, or if, we respond rationally and quickly enough to change our Titanic-like course.

ABOUT THE AUTHOR

Frosty Wooldridge graduated from Michigan State University. He is an environmentalist, mountain climber, Scuba diver, dancer, skier, writer, speaker and photographer. He has taught at the elementary, high school and college levels. He bicycled 100,000 miles on six continents and six times across the United States. His feature articles have appeared in national and international magazines for 30 years. He has interviewed on NBC, CBS, ABC, CNN, FOX and 150 radio shows. He writes bi-weekly columns for 40 web sites including www.NewsWithViews.com ; www.AmericanChronicle.com; www.neighbors.DenverPost.com ; www.examiner.com . He is the author of *Handbook for Touring Bicyclists; Strike Three! Take Your Base; An Extreme Encounter: Antarctica; Bicycling Around the World: Tire Tracks for your Imagination; Motorcycle Adventure to Alaska: Into the Wind; Bicycling The Continental Divide: Slice of Heaven, Taste of Hell; Immigration's Unarmed Invasion: Deadly Consequences.* He presents a program to conferences and colleges across the USA: "*The Coming Population Crisis in America: and what you can do about it.*" He lives in Louisville, CO www.frostywooldridge.com

PRAISE FOR—*AMERICA ON THE BRINK: THE NEXT ADDED 100 MILLION AMERICANS*

"The environmental community may be outrageously AWOL on the important subject of population, but not Frosty Wooldridge. Read *America on the Brink: The Next Added 100 Million Americans!*" Richard D. Lamm, Governor of Colorado 1975- 1987

"This is a veritable cannonade of a book. Wooldridge targets the people and institutions, from the President on down, who have refused to look at the consequences of population growth in the modern era. His focus is on the United States, but his range is the world. He describes most of the identifiable consequences of population growth. One useful feature of the book is the extensive use of quotations from experts who describe the problems we face and the price we pay for our inaction. He is fearless in taking on issues that politicians fear to mention, such as the effects of mass immigration on our population future and our social system. He closes with recommendations as to what can be done. They boil down to learn the facts and hammer your elected representatives with the demand that something be done about them. I commend this book to the young and to others still drifting without a sense of purpose. It might well persuade them that the best use of their energies would be to engage in the charge to force population issues into our national and local political decisions." Lindsey Grant, Writer and former Deputy Assistant Secretary of State for Environment and Population.

"Eye-opening, incisive and brilliant! The US has the fastest growing population of any industrial nation, and one of the world's highest per capita consumption rates. But we are only beginning to awaken to the natural limits on the resources that must support all of us voracious consumers. Water, topsoil, forests, fish, petroleum...it takes a lot to satisfy an American, and the more of us there are, the more pressure we exert on our environment. Many discuss our personal consumption patterns, but few dare talk about the underlying crisis of population growth. Wooldridge is one of the few courageous voices warning us about the implications of our current direction, and informing us what we can do to change course." Richard Heinberg, Author of *Peak Everything*, Senior Fellow, Post Carbon Institute

"Wooldridge's powerful indictment of our political leaders for failing to address U.S. overpopulation should be required reading in classrooms and boardrooms. To mindlessly add 100 million more people to our nation's population in just the next 40 years surely will put the United States economy and social order at further risk. Wooldridge carefully documents the consequences of what he calls "a human Katrina." His challenge needs to be addressed on national, international and individual levels before it is too late." William B. Dickinson, The Biocentric Institute (italics)

"In a world strangling on population overshoot, America has failed to take the one humane measure that might really avail to help us in our part of this struggle: enforcement of reasonable immigration laws. I hope this book will focus the public's attention on the issue." James Howard Kunstler, Author of *The Long Emergency*.

"Unlimited population growth is a more sinister threat to the welfare of humanity than global warming, pollution, plague, or starvation. These are but mere symptoms. The underlying cause of the upcoming catastrophe is overpopulation. Most people are unable or unwilling to face the truth, and will continue until the very sad and painful end of life as we know it. Overpopulation is a fact, so is the outcome. Wooldridge writes with clarity and balance only dreamed of my most modern writers. His work will stand as a beacon of sanity in a world gone mad." Kenneth R. Hampshire President, Syntratech Corporation

"So often when I hear, 'Overpopulation is the real problem', my many years in the field tell me that neither the speaker nor listener know the true meaning of those words. In his *Next Added 100 Million Americans,* Wooldridge gets into hundreds of nooks and crannies to show the exponentially escalating damage to our earth and to our lifestyles that flow from our ever increasing numbers. Get this book and then follow his recommendations for what you, your friends, and other individuals can do! Wooldridge makes a compelling case for all humanity." John R. Bermingham, Founder, Colorado Population Coalition.

"In my lifetime, one factor—more than any other thing--has changed the landscape, environmentally, socially, economically, of the United States: a population explosion. Subdivisions carpet mountain valleys in the previously "wide open spaces" of the American West, the American Southwest is in a population-driven water emergency, wildlife are in crisis or disappeared, while rapidly expanding human numbers in, appallingly, what is the world's third most populated and fourth fastest growing nation, no longer enrich our lives but denigrate them. Gridlock spreads beyond just freeways to hospital emergency rooms, schools, national parks and and the increasingly crowded and stressed everyday lives of all Americans. Yet, the United States mostly immigration-driven population explosion is the unacknowledged elephant in the room--including as pertains to global warming--by politicians, Big Media and others. *AMERICA ON THE BRINK* calmly but irrefutably argues, "Enough!" It is a must-read for anyone who treasures the hope of a promising future--rather than China-like overcrowding—for their children and their nation!" Kathleene Parker, Environmental activist/journalist Albuquerque, New Mexico

"Wooldridge's work is definitely not a book to curl up on the couch with on a rainy afternoon. His topic will most likely keep your attention late into the night. There are two kinds of people who seek knowledge: those who want to know the facts and those who are more comfortable listening to the "experts" in the field. As a testimony to Wooldridge's diligent efforts, I highly recommend this book. Do your own research and study. This book is a must read to all those who seek information and facts. This book involves important information and exposes the truth that the political elite don't want you to know about hyper-population growth." Jan Herron, journalist, Magic City Morning Star

"*AMERICA ON THE BRINK* asks readers to help reverse America's mindless march toward a population of one billion before our grandchildren and their children inhabit a nation that looks more like present day China than present day America. Here are the facts: America's population growth is perpetual—107 of the past 108 years. America's population growth is accelerating—76 million during the first half of the 20th century, but 130 million during the second half. Wooldridge's book amplifies these facts, explains how they came to be, and asks readers to help reverse America's march toward a population of one billion." Edward C. Hartman, Author of *The Population Fix: Breaking America's Addiction to Population Growth*

"I have been at work with many organizations since 1965, which seek to help bring about world population stability; my sense is that any form of "soft landing" (i.e. sans wider starvation, wars, etc.) is becoming less likely all the time. Wooldridge has put this case as completely, succinctly and powerfully as possible. The failed Bush Administration ideologically foreclosed the urgent action required on such issues as reproductive health, choice, and adequate contraceptive programs overseas. Thus as Wooldridge explains, the solutions which could have been so readily reached remain tenuous and the survival of an America as we know it and even our fragile planet put in great jeopardy." Donald A. Collins, President of International Services Assistance Fund, was a founding board member of the Guttmacher Institute, Ipas, and Family Health International and served on the National Board of Planned Parenthood of America

"For the better part of this decade, Frosty Wooldridge, writing with compelling prose, has been acting as the proverbial canary in the mine sending out warning signals regarding the deleterious effects of overpopulation and the environment. My suggestion is that you buy a copy of his book and send it to each of your national representatives. A clarion call in the form of a book! He writes brilliantly as to our future and what to do about it." Dr. John Copenhaver

"In a very readable format, it's all here. Wooldridge discusses population growth numbers, myths of growth, and growth's social and economic consequences, impacts on natural areas, the environment, and resources. Importantly, he provides sensible recommendations for change. Americans will do well to buy this book." Dell Eriksson, Director of Research, Minnesotans for Sustainability, energy and population researcher, author, and environmental activist.

"*AMERICA ON THE BRINK* by Frosty Wooldridge is an outstanding source of factual information that is very clear and concise. On overpopulation issues in the United States, I see a grim future for all citizens as well as lawfully admitted or naturalized immigrants. Politicians of both political administrations have advocated open borders and supported the massive surge of undocumented foreign nationals by the millions for "cheap labor "and illicit votes. U.S. Immigration laws ensure the lawful entry of all crossing the borders and excludes those that could pose a threat to the nation. Overpopulation as well as crime and impact on society are common sense goals that are rarely considered by elected leaders in their quest for a North American Union." John W. Slagle (ret) Anti-Smuggling Special Agent USBP

"Unbridled population growth poses the single greatest threat to our quality of life and ecosystems. Americans must recognize, admit and come to grips with this frightening reality or efforts toward a sustainable future will be futile. However, any rational discussion of over-population and its deleterious effects on all aspects of life in the U.S. is mysteriously absent from public discussion. Wooldridge convincingly shifts the current paradigm of unlimited population growth to one of population stabilization and the need for a national population policy. He presents a compelling case for reversing runaway population growth by addressing public and media apathy and denial with clear steps for solutions and citizen action. A must read for every American who cares about this country's future." Beth Thomas

"I've known Wooldridge as a passionate author, who loves America and alerts of the troubles that lie ahead. He warns government officials to take action. The book is an invitation to proactive ideas, putting together many minds to a common goal of solutions for a much greater and safer America and the neighborhood of nations where the American Dream is reachable to all." Juan José Herrera, journalist, Caracas, Venezuela

"Wooldridge cuts to the chase with this latest book: *AMERICA ON THE BRINK: THE NEXT ADDED 100 MILLION AMERICANS.* The facts are facts, the numbers are what they are and this book is an educational book. I recommend this book for everyone as America's recognition of this detrimental growth issue is needed soon. There are actions that we can take if we are informed and with many thanks to Wooldridge they are spelled out here for all to consider. If our country does not address the burgeoning population explosion in America, we will face the future of an unsustainable populace. Wooldridge's book offers clear solutions." Marty Lich-- researcher, writer, activist

"We ignore the implications of unchecked population growth and mass immigration at our own peril. When Third World migrants reach America, their ecological footprints balloon from near zero to first world levels of consumption, pollution, sprawl, and waste. If the demographers' predictions of an additional 100 million new Americans over the next thirty-five years come true, the result will be a veritable "Human Tsunami" upon the planet's ecology. This book will educate and guide you to become part of the solution." Michael Crowe, immigration reform activist

"When denial of an obvious truth borders on the hysterical, be very suspicious that cult-like forces are at work. How long do the environmental community and main-line media think they can get away with the lie that immigration-driven population growth is not at the heart of our most serious national problems? Frosty Wooldridge has always been fearless in identifying this dangerous myth and with this book he has done it again." Diana Hull, Ph.D. President Californians for Population Stabilization, Santa Barbara, CA

Printed in the United States
217157BV00003B/7/P